To my colleague Laura —
looking forward to
conversations about
naturalism & the
nature of inquiry.

Paul Davis

Norms of Nature

Norms of Nature
Naturalism and the Nature of Functions

Paul Sheldon Davies

A Bradford Book
The MIT Press
Cambridge, Massachusetts
London, England

This book was set in Sabon by Achorn Graphic Services, Inc., and was printed and bound in the United States of America.

Library of Congress Cataloging-in-Publication Data

Davies, Paul Sheldon.
Norms of nature: naturalism and the nature of functions / Paul Sheldon Davies.
p. cm.
"A Bradford book"
Includes bibliographical references (p).
ISBN 0-262-04187-1 (hc : alk. paper)
1. Natural selection. 2. Naturalism. I. Title. II. Series.
QH375.D36 2001
570′.1—dc21 00-055403

for Ann
my natural norm

Consider, anatomize the eye: Survey its structure and contrivance; and tell me, from your own feeling, if the idea of a contriver does not flow in upon you with a force like that of sensation.

—David Hume 1779, *Dialogues Concerning Natural Religion,* Part III

Organs of extreme perfection and complication—
To suppose that the eye, with all its inimitable contrivances for adjusting the focus to different distances, for admitting different amounts of light, and for the correction of spherical and chromatic aberration, could have been formed by natural selection, seems, I freely confess, absurd in the highest possible degree. Yet reason tells me, that if numerous gradations from a perfect and complex eye to one very imperfect and simple, each grade being useful to its possessor, can be shown to exist; if further, the eye does vary ever so slightly, and the variations be inherited, which is certainly the case; and if any variation or modification in the organ be ever useful to an animal under changing conditions of life, then the difficulty of believing that a perfect and complex eye could be formed by natural selection, though insuperable by our imagination, can hardly be considered real.

—Charles Darwin 1859, *On the Origin of Species,* Chapter VI

Contents

Preface

The old argument of design in nature, as given by Paley, which formerly seemed to me so conclusive, fails, now that the law of natural selection has been discovered. We can no longer argue that, for instance, the beautiful hinge of a bivalve shell must have been made by an intelligent being, like the hinge of a door by man. There seems to be no more design in the variability of organic beings and in the action of natural selection, than in the course which the wind blows. Everything in nature is the result of fixed laws.

—*The Autobiography of Charles Darwin* (Barlow 1958)

Darwin was right about many things. The argument from design is among the things he came to see aright. For the theory of evolution by natural selection explains the adaptedness among living things better than the argument from design. And it does so without recourse to a designer. Of course, the deity may have created the world according to fixed laws, including those involved in natural selection, but Darwin is making a point about scientific inquiry. Logical compatibility is not enough. A scientific theory is compelling when it provides explanatory resources that are relevant and fruitful, and when it leaves itself open to reasonable sorts of test. On these criteria, evolutionary theory is compelling but divine design is not. So the argument from design fails in competition with evolutionary theory. In the ecology of ideas, God has become extinct.

No doubt, few philosophers today are tempted to argue for the existence of God by appeal to natural design. But more than a few are curious to understand the apparent design possessed by living systems. The mammalian eye strikes us as a highly purposive natural device. The idea of design, if not that of an actual designer, strikes many of us with the force of sensation. This is a fact we wish to explain. We want to know what in the world makes it true that token eyes are purposive or, at minimum, what in the world inclines us to classify eyes and other natural traits as purposive.

Contemporary theories of functions claim to address this query. They assert that, while Darwin deprived God of any remaining scientific legitimacy, the living realm nevertheless exhibits a kind of design best explained in terms of natural selection. After all, selection sorts among variants of traits, and the less efficacious are selected against while the more efficacious are selected for and preserved. Selection thus favors those variants that better perform certain tasks and better satisfy selective demands. On this view, the selective efficacy of ancestral tokens endows descendent tokens with the function of producing the selectively successful effects. The function of the mammalian heart, for example, is to pump blood because pumping blood was selectively efficacious among ancestral mammals. Precisely this appeal to selective success is supposed to account for the evident purposiveness of natural traits.

The appeal to natural selection does not end there, however. Contemporary theories of functions aspire to account for a rather robust form of purposiveness. They aspire to explain malfunctions. The aim is to explain how it is possible that natural traits can fail to fulfill their functional task and nevertheless retain their functional standing. The assumption is that a diseased or damaged eye is malfunctional rather than nonfunctional. A diseased eye is "supposed to" enable the organism to see even though the requisite capacities are absent. Functional properties, on this view, are normative; they are equivalent to norms of performance imposed by ancestral selective success. The crucial idea is that a token eye malfunctions when it fails to perform the task that ancestral tokens performed with selective efficacy. The token remains in its functional category thanks to its selective history—whether it can ever perform the functional task is inessential. Precisely this appeal to selective success is thought to make possible the occurrence of malfunctions. And all of this is said to help us understand, at least in part, the design among natural systems.

A central aim of this book, however, is to convince you that we have been going the wrong way in our attempts to understand the norms of nature. I shall argue that, while most natural traits indeed have functional properties, those properties are not constituted by the selective success of ancestral tokens. Natural selection, I believe, is inessential to the existence of functions. A further aim is to convince you that we already have an alternative theory of functions far more plausible than the currently popular historical view. I shall develop this alternative in ways that, to the best of my knowledge, are novel and promising. And I shall be especially concerned to show that this alternative fits our emerging naturalistic worldview better than the historical approach to functions.

The book consists of two quite general lines of argument. I argue that we do better to conceptualize functions as effects of systemic components that contribute to more-general capacities of the larger system. Emphasis falls upon the structure and organization of natural systems and the work accomplished by components of such systems. The function of the mammalian heart is to pump blood because pumping contributes in significant ways to certain capacities of the circulatory system. Functions are essentially systemic; history is not essential. Functions typically have a history, to be sure, including a selective history, but pumping would have been the function of the heart even if some other device had beaten it out early in the evolution of mammals. Functions are contributions to systemic capacities and, while selection can preserve or eliminate those functions, selection is not their source.

I also argue directly against the attempt to understand functions in terms of ancestral selective success. Several arguments support my negative assessment, but one rather central reason for rejecting the historical approach is its retention of the notion of design. The appeal to design may be unintentional, but it is there all the same, dressed up in claims about the possibility of malfunctions. Everyone agrees that Darwin's discovery defeats the argument from design; that much is common ground. But advocates of the historical approach also assert that living systems nevertheless have a kind of design. What is meant, of course, is that traits endowed with historical functions now possess functional roles that persist even when the requisite physical capacities are lost due to damage or disease. These roles underwrite the attribution of a kind of design insofar as there is something that descendent tokens are "supposed to" do even when they are incapacitated. And precisely this attribution of design appears to conflict with Darwin's claim that there is no more design in selective success than in the blowing of wind.

Advocates of the historical approach might retort that it makes perfectly good sense to talk of natural design so long as we refer only to the evident *adaptedness* of living things and eschew any sort of theology. And they also might insist that Darwin himself asserts that natural traits are adapted and hence "designed" in this restricted sense of the term. But this, I shall argue, is disingenuous. We can all agree that some organisms or organismic traits are better adapted than others. It is quite another matter to assert that there are norms of nature imposed by ancestral selective success, norms that persist in the face of physical incapacitation, norms that make possible the occurrence of malfunctions. The postulation of such norms gives contemporary theories of functions their initial intuitive appeal; it appears to explain our sense that living things are purposive. But it also imputes to living things a kind of design that far exceeds mere adaptedness. Mere adaptedness, viewed from a naturalistic perspective, provides no grounds for the postulation of any such norms. This is clear in the natural selective explanations offered in Darwin's work. There are claims of adaptedness, to be sure, but no appeals to such norms. The postulation of such norms puts contemporary theories of functions in conflict with the theory of evolution by natural selection. Or so I shall argue.

Darwin rejected the argument from design because it made for inferior science, because he had a theory that better accounts for the phenomena. We should reject the attempt to understand functions in terms of natural selection on analogous grounds. Understanding functions in terms of selection commits us in ways that conflict with the methods and postulations of our best natural sciences. And anyway we have an alternative theory that better accounts for the phenomena—the theory developed in this book.

Norms of Nature

1 Tracing Links of Causation

Much in the biological world appears to us purposive. Much appears governed by quite specific norms. We use these norms to assess the performance of certain objects, to determine whether or not they have fulfilled their functional roles, whether or not they have done what, in some sense, they are supposed to do. It is natural to think that the eyes of mammals are for seeing, that the wings of birds are for flying, that incisors are for tearing and molars for mashing, and so on. The theoretically inclined thus will wonder, Are there norms of nature? If so, what are these norms? From whence do they arise? How do they attach to just some traits and not others? If there are no such norms, why are we so inclined to see some parts of nature—eyes, wings, teeth—as purposive and other parts—rocks, clouds, moons—as nonpurposive? And whence this inclination? This book ventures answers to these questions.

These questions concern a striking feature of life. Living things change and virtually all such changes appear purposive. The appearance of purposiveness in the biological realm is as ubiquitous as it is striking. Organisms develop and mature, recover from illness and injury, adjust to changes in environment, and engage in myriad behaviors. Organisms also suffer disease, decay, and death, and even these processes can appear purposive. It thus is plausible, perhaps natural, to conceptualize change among the living as purposive. Anyone striving to understand the world we inhabit will be drawn to questions about these phenomena. However, in addition to their intrinsic interest, these questions arise from a prima facie tension between the methods and postulations of the natural sciences generally and the attribution of natural norms. We tend to believe that physical phenomena are the results of natural laws governing physical elements. We tend to think of physical theories as accounts of the ways things are and the way things behave, and perhaps also the ways things might have been. But we tend to think differently about biological phenomena. Our biological theories account for the way things are, to be sure, but they also account for the way things should be. We tend to think, for example, that a malfunctioning eye, despite its defective condition, nevertheless is supposed to perform its functional task. There is a norm of performance that applies to eyes and persists in the face of incapacitation. Or so we tend to believe. The philosophical question, then, is, What are these norms of nature such that we appeal to them in the course of inquiry and such that they do not conflict with the methods and postulations of the natural sciences generally?

Biologists appeal to functions in the course of theorizing about the biological realm and it might be thought that we should turn to them for an account of what functions really are. It might be thought that the explicit or implicit understanding of functions among biologists will answer the above question. I have doubts about that. The task is to show how the notion of functions in biology fits within the larger framework of the natural sciences generally. It seems naive to think that biologists, in their appeals to functions, are concerned to solve this problem. It is one thing to appeal to functions in the course of theorizing; it is quite another to theorize about such appeals and the extent to which they cohere with other areas of inquiry. Of course, theoretically inclined biologists may explain how functions fit the methods and postulations of the natural sciences generally, but there is nothing about being a biologist that provides especial qualifications for this sort of job. Nor is there good reason to think that biologists, as opposed to nonbiologists, have purged from their understanding of functions all the metaphysical—including the theological—strains that infect the history of this notion. One can, after all, be a superb biologist and a committed theist or deist, as several of Darwin's peers demonstrated.[1] Surely one can be a fine biologist and nevertheless employ a notion of functions that does not fit comfortably the postulations or methods of the natural sciences generally. So the job,

1. Among the theists were Sedgwick and Whewell; among the deists were Herschel and Lyell. See Ruse (1979), chapter 3 and Gillespie (1979), chapter 5. See also chapter 5, section III below.

as I see it, is not to explicate or describe the way that we or certain specialists employ the concept of functions. The job, rather, is constructive and, if need be, revisionary. The job is to suggest how an apparently powerful concept is best understood within a broad domain of inquiry.[2]

Contemporary theories of the norms of nature fall into three general categories, each of which I describe fully in chapter 2. (1) The *historical approach* asserts that functions are the product of some type of historical process. The most prominent version of the historical approach asserts that the functions of any trait emerge out of the selective success of ancestral tokens of that trait. Hearts have the function of pumping blood because pumping blood is the effect of ancestral hearts that contributed to selective success and thus caused the persistence of hearts. Advocates of this approach tend to assume that some natural traits are genuinely functional while others are merely useful; they assume that the effects of some traits are "proper" to those traits, while the effects of others are incidental. The main aim of the historical approach is to account for the alleged "properness" of functions in terms of some sort of historical success. (2) The most prominent alternative to the historical approach is the *systemic capacity approach.* On this view, functions are attributed relative to the larger system within which the trait operates and relative to certain capacities of the larger system. The functions of any trait are those effects that, within the context of the system, contribute to the exercise of some higher-level capacity. Hearts have the function of pumping blood, relative to the capacity of the circulatory system to distribute nutrients throughout the body, because pumping blood contributes to the exercise of this higher-level capacity. Critics of this approach charge that it is incapable

2. In this respect, my approach differs from that of most contemporary theorists. From Brandon (1981, 1990), through Neander (1991) and Griffiths (1993), and up to Walsh (1996), Preston (1998), and Buller (1998), the approach to functions is generally conservative. The approach is to conserve what appear to be central intuitions concerning the nature of functions—intuitions of biologists and other specialists, mainly. The view defended here, by contrast, advocates the revision of at least one feature of functions thought to be conceptually central—the alleged normativity of functions that underwrites the attribution of malfunctions. On my view, the functions of natural, nonengineered traits are not (and cannot be) normative in the way most theorists suggest, and I offer an alternative approach to understanding malfunctions.

of distinguishing between effects of traits that are "proper" and those that are incidentally useful. In consequence, it is generally believed that, even if the systemic approach is compelling in its own right, an additional theory of functions is required to account for genuinely teleological effects. And that brings us to (3) the *combination approach,* consisting of various attempts to combine the historical and systemic capacity approaches into one theory. Advocates of this approach tend to assume that historical functions—unlike systemic functions—account for effects that are genuinely teleological. Motivation for the combination approach derives from this assumption, combined with the claim that systemic functions nevertheless serve important theoretical aims. One version of the combination approach claims that there is a single, unified theory that subsumes both views under the concept "design." Another holds that both views of functions are required but must be kept separate on the grounds that each applies to distinct phenomena. And additional options have been suggested.

My aim is to develop and defend a distinct version of the second approach. In its original formulation (Cummins 1975), the theory of systemic capacity functions asserts that functions are causal or structural capacities that contribute to the exercise of some larger systemic capacity within a complex system. I accept this basic view. In addition, I defend four substantive theses, each of which extends or revises the original theory. The first asserts that the systemic approach is more general in scope and, in fact, subsumes the historical approach. The systemic approach attributes functions to the components of populations and organisms affected by selection, as well as systems not affected by selection. This approach thus warrants all the functions attributed from within the historical approach, and numerous others as well. The historical approach, in consequence, taken as a separate autonomous theory, is redundant and ought to be discarded. I defend this claim in chapter 3. The second thesis is that the systemic approach ought to be restricted to systems that are hierarchically organized. This restriction blocks the oft-repeated criticism that the systemic approach is promiscuous in the functions it ascribes. I defend this claim in chapter 4. The third thesis concerns the intuition we have that certain features of some traits are genuinely functional, while others are useful but not "properly" functional. It also concerns the related intuition that certain traits, when damaged or diseased, are properly classified as malfunctional, while other traits, upon losing their utility, lose their functional status. My thesis is that, while both intuitions are in error, we nevertheless can explain the source of these intuitions. My Humean suggestion appeals to the effects of highly regular or highly complex hierarchical systems on our psychology. I speculate that such systems, insofar as they are self-preserving or self-perpetuating, cause us to expect their components to continue on in the same way, thereby generating the intuition that only some things are functional and that only some things qualify as malfunctional. I defend this revisionist claim in chapters 5, 6, and 7. The fourth thesis is that the systemic approach has superior naturalistic credentials. In particular, the attribution of systemic capacity functions is open to empirical and theoretical tests and, moreover, the theory as a whole is confirmed by the detailed case studies offered, for example, by Enç (1979) and by Bechtel and Richardson (1993). I defend this claim in chapter 6.

Virtues of the systemic capacity approach are highlighted by the defects of the historical approach. The first defect, noted above, is that the historical approach is redundant on the systemic capacity approach and hence ought to be discarded. The second is that the historical approach, in attempting to account for the possibility of malfunctions, commits itself to the existence of quite specific norms of performance that are noncausal and nonphysical in nature. I defend this construal in chapter 5. I then argue that we have powerful naturalistic reasons, external to the theory of selected functions, to reject the postulation of such norms. Third, and contrary to the claims of its advocates, the historical approach also lacks the internal resources with which to account for the possibility of malfunctions. I defend this claim in chapter 7. Accounting for malfunctions is said to be among the central virtues of the historical approach. So if the arguments of chapters 5 and 7 are sound, we ought to conclude that the historical approach, by its own lights, is a failed theory. Moreover, insofar as the combination approach inherits the central substantive theses of the historical approach, the defects of the latter infect the former.

Cast generally, then, I advocate the view that functions are nothing more than systemic capacities that contribute to the exercise of higher-level capacities we wish to understand and control. Functions, on this view, are effects that play a role in the workings of hierarchical systems. This is to emphasize the mundane but central fact that functional properties concern the operations or workings of those parts of nature that are systemic in a rather minimal sense of the term. The relevant sorts of systems are numerous and diverse. We may analyze the capacity of salt molecules to dissolve in water in terms of the systemic functions of constituent molecules. Or we may analyze the capacity of the circulatory system to distribute nutrients in terms of the systemic functions of its components. Significantly, we can also conceptualize entire populations of organisms as systems that accomplish various kinds of work. Populations evolve in certain ways; populations also remain in equilibrium in certain ways. We thus may analyze the capacity of a population in terms of the systemic functions of its structural components. Conceptualizing functions in this way—in terms of systemic work—enables the theory to attribute systemic functions to components of even the simplest systems, as well as systems of great complexity. This, as I argue in chapter 3, is an important result, as it dissolves the need for two separate theories of functions—selected and systemic—and enables us to formulate a single, comprehensive version of the theory of systemic functions.

My answer, then, to the question, Are there norms of nature? is the following: There are components of natural systems endowed with capacities of great importance to the workings of the system. The aim of scientific inquiry quite generally is to discover, understand, and control those systemic capacities. But the components of natural, nonengineered systems, contrary to the claims of the historical approach and contrary to most versions of the combination approach, possess no norms of performance. They possess systemic capacities the effects of which enable the larger system to work in ways we wish to understand. But they do not possess the sorts of properties attributed from within the historical approach to functions. Several of the arguments developed throughout this book are intended to establish this negative view. The truth of this view, however, leaves ample room for an array of other sorts of norms, many of which are aptly characterized as natural. These are norms that emerge in the course of various human activities. There are, for example, epistemic norms generated by our informed expectations of certain categories of traits classified in terms of systemic capacities. These are important, I believe, in understanding why we are so inclined to classify incapacitated token traits as malfunctional. I return to this topic in chapters 6 and 7. My view of the norms of nature also leaves room for norms that emerge in the course of personal, social, legal, commercial, and ethical relationships.[3] And my answer to the question, Why are we inclined to see some things as functional and others as nonfunctional? is that certain sorts of systems cause us to expect effects that, relative to the systemic capacity we wish to understand, play especially salient systemic roles. This will become clearer in the course of my discussion of epistemic norms and the attribution of malfunctions.

Whatever the full range of natural norms involved in various human activities, my concern here is first and foremost with the sorts of norms thought to apply to nonengineered natural traits at all levels of biological organization, from the molecular to the behavioral. My concern is with the norms said to apply to genes and neurotransmitters, to hearts, eyes, and teeth, to foraging strategies and mating displays. These traits, it is claimed, are purposive or functional quite apart from our epistemic or practical concerns; these traits possess norms of performance quite apart from our attempts to understand or control them. The claim is that these norms have arisen from historical-causal-mechanical processes, and nothing else. And it is this claim I shall be concerned to assess, for I believe it to be mistaken.

The focus of my discussion thus excludes the functional status of artifacts. The functions of artifacts involve the intentions of designers and manufacturers and the social, legal, and ethical relations between manufacturers and consumers. An adequate account of artifact functions must explain how the relevant intentions and conventions help produce and sustain such functions. But these factors do not apply to the functional standing of nonengineered natural traits. It is naive to think that the functions of nonengineered natural traits—the workings of which are causal-historical and independent of our intentions or conventions—suffice as a model for understanding the functions of artifacts. It is equally naive to think that the functions of artifacts—the workings of which depend

3. I have in mind the views of Blackburn (1998), Gibbard (1990), and Harris (1999).

upon our intentions and conventions—provide an adequate model for understanding the functions of natural traits. The theory of systemic functions, as developed here, is intended to apply foremost to nonengineered natural traits. In consequence, I shall appeal to the functions of artifacts sparingly and with caution. Extending the theory of systemic functions to the relevant sorts of intentional and conventional factors involved in the functions of artifacts is a project for some other occasion.

The view I defend is also revisionist in nature. It aims to dispense with the notion that natural traits are "properly" functional or "for the sake" of some end. Function attributions are important in inquiry, as I intend to show, but their importance in no way requires us to believe in functions that are "proper" or otherwise distinct from other sorts of systemic capacities. Such revisionism might give the impression that I endorse what Enç and Adams (1992) call the "eliminativist view" of functions, according to which there is no difference between dispositional properties generally and functional properties specifically:

> [Eliminativist] views identify functions merely with the activities (or dispositions) of a character that happen to interest the investigator, and they effectively reject the intuition that a real difference exists between dispositions and functions. (Enç and Adams 1992, 637)

Enç and Adams reject the eliminativist option, claiming that it conflicts with common sense. This is to assume that a condition of adequacy on any theory of functions is that it distinguish functions from dispositions generally. This seems a plausible assumption—though I think we should be prepared to discover that no such distinction can be sustained and that eliminativism is the correct view. Nevertheless, my view does not belong in the eliminativist camp; my view is revisionist without being eliminativist. The theory of systemic functions dispenses with the postulation of properties intended to explicate our sense that some effects are "proper" or "for the sake" of some end. Hence the revision. But the theory asserts that functions are specific capacities within certain kinds of systems relative to certain systemic capacities of the larger system. Not just any kind of disposition qualifies as a systemic function. As I argue in chapter 4, systemic functions arise only in the context of hierarchical systems and only when they contribute to a real capacity of the larger system. Our explanatory interests may be important in the discovery of systemic functions, but our interests are neither necessary nor sufficient for the existence of such functions. And that makes my version of the theory revisionist without being eliminativist.

Portraying and defending the above view is the task of the chapters that follow, but the general picture can be sketched in broad strokes from the thoughts of Aristotle, Hume, and Darwin, all of whom puzzled at length over the apparent purposiveness of nature. Consider Aristotle's introductory remarks to *Parts of Animals*. Studying animals, he says, no matter how grotesque, gives us amazing pleasure:

> Having already treated of the celestial world, as far as our conjectures could reach, we proceed to treat of animals, without omitting, to the best of our ability, any member of the kingdom, however ignoble. For if some have no graces to charm the senses, yet nature, which fashioned them, gives amazing pleasure in their study to all who can trace links of causation, and are inclined to philosophy. (*Parts of Animals,* 645a, 5–10)

The pleasure comes from tracing links of causation that operate within animals. But not just any links will do. Animals, according to Aristotle, are "fashioned" by nature. We experience pleasure by tracing links that reveal the way in which an animal's parts are "put together," by uncovering links the efficacy of which fulfills the proper "ends" of the animal. In doing so, we encounter a form of the beautiful:

> Absence of haphazard and conduciveness of everything to an end are to be found in nature's works in the highest degree, and the end for which those works are put together and produced is a form of the beautiful. (645a, 23–25)

Now, the parts of animals can elicit an aesthetic response. Evidence for this is compelling. Many capacities of organisms involve complex interactions of several parts and layers of functional dependence, and understanding and appreciating the exercise of such capacities—respiration, circulation, digestion, locomotion, thought, speech, and more—affects us aesthetically. Indeed, I read Cleanthes's remarks in the opening epigraph as expressing one such aesthetic reaction (more on this shortly). Moreover, animal parts tend to serve animal needs, as traits that fail to contribute to an organism's survival or reproduction tend to atrophy or disappear thanks to selection and regressive evolution. It is not the case, however, that animals or natural systems generally are "put together" or "produced." On a more contemporary view, organisms evolve and give rise to further forms of life, but no one and no thing fashions the forms that emerge. They emerge thanks to the effects of evolutionary forces—including selection, drift, and migration—that preserve or eliminate forms that arise by mutation and recombination. Of course, selection has the effect of sorting the efficacious from the nonefficacious or the less efficacious and, as Ayala (1970) points out, this type of sorting, repeated over stretches of evolutionary time, can give rise to complex, adapted systems that otherwise would not have evolved. But selection, like other evolutionary processes, depends upon factors that are blind, nondirectional, and, to use Aristotle's term, haphazard: Selection depends upon the range of actual mutations, the extent of migration, and the contingencies of the selective regimes. And although selection may have the effect of sorting from among such materials, that hardly amounts to the "crafting" or "putting together" of such systems. The pleasure we enjoy in tracing causal links, then, contrary to Aristotle, cannot come from discovering the ends for which an animal is put together, for the realm of nonengineered natural traits is devoid of such ends.[4]

Since the ends to which Aristotle appeals are illusory, so too are the Aristotelean functions ascribed to the parts of animals. I accept, however, Aristotle's claim that tracing the operations of hierarchical systems such as animals is both important and attractive. In tracing links of causation that give rise to more-general systemic operations, we discover the systemic functions of natural objects. This, as we shall see, is important in various theoretical endeavors. Moreover, there is pleasure in tracing these links. We discover forms that seize our attention, that fascinate and de-

4. Ayala runs together language describing the production of artifacts and language describing the causal-mechanical processes that give rise to nonengineered natural traits: "The hand of man is made for grasping, and his eye for vision. Tools and certain types of machines made by man are teleological in this sense" (Ayala 1970, 9). Tools and machines are for certain ends, to be sure, but it is doubtful that the same is true of hands and eyes. We certainly should not take this as a datum, as part of the natural phenomena; rather, it should be treated as a conclusion in need of argument. As I point out above, the intentional and conventional features of artifactual functions do not apply to natural functions. Hands and eyes have an evolutionary history, of course, but they were not made for the performance of some specific task.

light us, that perplex and disgust us, that fill us with awe. We are natural systems that, thanks to our cognitive, volitional, and sensory systems, are capable of being moved in these ways by the natural systems with which we interact. It would be a great puzzle if things were otherwise. It would be puzzling if the living things with which we interact failed to command our attention and provoke discernible aesthetic reactions in us. It would be puzzling if we did not inherit from our ancestors the propensity to be gripped by living things all around us. We would be left to wonder how our species has managed to survive.

My reaction to the historical approach to functions parallels my reaction to Aristotle. Advocates of the historical approach posit functional roles with associated norms of performance. These roles are said to persist even when the requisite physical capacities are lost. But we should reject these norms as surely as we reject Aristotelean ends; this, at any rate, is the thesis of chapters 5 and 7. At the same time, and as I suggest in chapter 3, we can understand and preserve the nonnormative elements of the historical approach from within the systemic capacity approach. On my view, the functions that emerge as a consequence of ancestral selective success are nothing more than one kind of systemic capacity function. This kind of systemic function emerges in the context of an analysis of the population and the population's components, including an analysis of the varying organisms and their varying capacities. We thus can trace theoretically important links of causation within a population and identify salient systemic capacities. And we can do so without positing the existence of norms that persist in the absence of requisite physical mechanisms.

The opening epigraph from Hume is a key passage from Cleanthes's "irregular" argument for the existence of the Judeo-Christian God. The argument asserts an analogy between the natural and the artifactual realms, but makes the further claim that our knowledge of or access to this analogy is noncognitive. The claim, at minimum, is that our grasp of the similarity between the natural and the artifactual is not exclusively cognitive—the idea of a contriver hits us like a sensation. If we look without prejudice upon the complex adaptations of the world—the eye, for example—we simply see that such objects are the products of an intelligent contriver. We do not infer it; we see it. Or, at the very least, what we see causes us to feel that such objects must be the result of intelligent design; if we do not literally see it, we at least feel it. This claim has an obvious Humean flavor, appealing to a sentiment that appears pervasive and entrenched. This claim is also surprisingly modern in flavor. It is similar in spirit to the intuitions, described by most contemporary theorists of functions, concerning the normative status of functional properties. Cleanthes thinks that marks of intelligence are there on the surface. Similarly, contemporary theorists take it for granted that there are marks of design in the natural realm and that nature is designed even though there is no designer.

At any rate, it is striking that Philo, Hume's skeptic, does not endeavor to refute this version of the argument from design. This is striking given Hume's general propensity to explain away recalcitrant phenomena—miracles, freedom of will, causal necessity—in psychological or sociological terms. An appeal to our psychological capacities would be especially attractive after Part VIII of the *Dialogues,* where Philo develops the Epicurean chance hypothesis concerning the origins of natural order. The chance hypothesis provides a how-possibly explanation for the emergence of order that in no way appeals to an intelligent designer. Having shown that order could arise from purely causal-mechanical processes, it is plausible to suggest that our inclination to see the world in terms of design is explicable in light of certain facts about our psychology, rather than facts about the nonpsychological world.

The theory of systemic capacity functions, as developed below, adopts this Humean strategy in limited fashion. I appeal to various psychological speculations to display the relative strength of the systemic functions approach and also to explain away some of the presumed phenomena that motivate the historical approach. As we have seen, some advocates of the historical approach contend that some natural things—eyes, wings, teeth—possess genuine norms of performance, while other things—rocks, clouds, moons—may function as this or that but possess no such norms. Some things are purposive, others merely useful. Advocates of the historical approach assert that they can account for this difference while the theory of systemic functions cannot. But this is not so. As I argue in chapters 5 and 7, there are good reasons for rejecting the claim that natural objects such as eyes or teeth possess the sorts of norms attributed from within the historical approach. Moreover, and as I argue in chapters 4 and 6, the theory of systemic functions can account for the fact that we are inclined to see some things as more functional than others. The theory accounts for this inclination in good Humean fashion, without countenancing the roles and norms of performance posited within the historical approach.

Finally, the second epigraph with which this book opens comes from a well-known section of the *Origin* in which Darwin admits that complex adaptations like the eye pose a powerful prima facie challenge to his theory. But he also insists that, if guided by reason, we can construct a plausible how-possibly explanation of the evolution of such adaptations, no matter how complex they may be. Like Hume's Philo, Darwin is concerned to establish the possibility that order and adaptations arise from purely causal-mechanical processes. But unlike Philo (and Hume), Darwin offers a specific kind of causal mechanism and an abundance of evidence for the existence and efficacy of that mechanism; Darwin's how-possibly explanation, once filled out, shades into a powerful how-actually explanation.[5] In defending the claim that the order and complexity of adaptations like the eye could have arisen as a gradual and cumulative effect of natural selection, Darwin is resisting the creationist tendencies of his peers. As Hull (1973) demonstrates, Darwin's critics—including his most accomplished peers—could not accept the suggestion that natural selection is potent enough to explain the emergence of traits as functional as the eye. Selection alone, they claimed, could not explain the emergence of such highly adapted complexity; some form of intelligence was required, if only to guide the evolutionary process from afar. And as Gillespie (1979) shows, the doctrine of creation was embraced in various forms. Some theorists (for example, Agassiz) insisted that new species had to be the effects of special acts of creation; others (Richard Owen) insisted that the emergence of new species had to be ordained by a predetermining will; still others (Asa Gray, Charles Lyell, and Alfred

5. Brandon (1990) appeals to the notion of a how-possibly explanation in order to characterize the empirical constraints on adequate selective explanations. The considerations offered address the worry, pressed by Gould and Lewontin (1979), that adaptationist explanations all too often are illegitimately unconstrained. See Brandon (1990), chapter 5.

Wallace) insisted that the variation necessary for selection had to be the work of an intelligent, designing deity. These critics agreed that God's designing intentions enter the historical process at some time or other. They agreed that our explanations had to appeal to God's intentions because the design of the natural world was simply too complex and too vast—marked by too much intelligence—for natural selection.

Advocates of the historical approach to functions, of course, do not embrace theism or deism—not, at any rate, in their philosophical works. Unlike Darwin's critics, they insist that selection can account for natural design. Nor do they claim that nature is literally designed, only that it has a kind of design, one that arises out of the process of selection: ". . . one of Darwin's important discoveries is that we can think of design without a designer" (Kitcher 1993, 380).[6] But it is here, in the postulation of natural design, that I demur. Advocates of the historical approach and of the combination approach assert that the biological realm is designed in some respect or other and that we can understand this design in terms of natural selection. On my view, it is a mistake to hold that nature is designed in any sense of the term. There is order, regularity, degrees of complexity, and degrees of adaptedness—but no design. Of course, we analyze natural systems into systemic capacities in order to understand and control their operations, and it is tempting to conceive of these systems in terms of some sort of design. But we are wrong to give in to such temptation. Natural systems are comprised of parts that interact with one another, and sometimes these interactions are astonishingly complex and elegant. But that shows only that natural systems work—they exercise higher-level capacities by virtue of organized lower-level capacities—without having been designed and without exemplifying marks of design. Darwin did not show us how to understand the world in terms of design despite the absence of a designer; he showed us instead that we ought to stop thinking of the world in terms of design. He showed us that the biological realm exhibits great regularity and complexity as a result of nothing more than causal and mechanical historical processes.

6. This type of claim is surprisingly common in the literature on functions, from Ayala (1970) to Allen and Bekoff (1995). And other philosophers appeal to this claim in theorizing about various phenomena. Dretske (1995), for example, endorses Kitcher's assertion in the course of developing a theory of consciousness.

My aim is to show how to understand the attribution of functions in biology and elsewhere in a way that coheres with the methods and postulations of the natural sciences generally. And my strategy is to develop a theory of functions compatible with the fact that the nonengineered, natural realm is devoid of design. To accomplish this, we must revise—not simply explicate—the concepts involved, especially our concepts of function and design. We must relinquish the associations we have between these notions and the norms of performance posited by the historical approach. Our naturalistic scruples demand this. On the view offered here, the living realm is populated with complex, adapted systems that work, that accomplish a seemingly endless array of tasks. All these systems are products of evolutionary processes. But none of these facts underwrites anything more than the attribution of systemic capacity functions. The urge to attribute norms that violate our naturalistic commitments is, I believe, best explained by considering the nature of the urge. We can trace the causal and structural links between the parts of plants and animals and thereby formulate predictions and explanations, but to think we can do more is to regress to a brand of metaphysics untethered from the world we strive to understand.

2

Approaching the Norms of Nature

The historical and the systemic capacity approaches to functions are taken by many to be the main contenders, though additional views have appeared recently. In particular, various attempts to combine elements of both views have been defended. The purpose of this chapter is to explicate the various versions of the combination approach as well as the historical and systemic approaches, and to indicate in general terms where my view differs. Development and defense of my view are reserved for subsequent chapters. My intent is to depict enough of the three main approaches and enough of the alternative I endorse to enter the argumentative fray in chapter 3.

I Historical Approaches

The most pervasive account of the norms of nature is the historical approach. On this view, the norms that apply to the performance of genes, hearts, and all the rest are the products of historical processes that resulted in the perpetuation of those kinds of traits. The most pervasive version of the historical approach appeals to success wrought by natural selection. Functional properties, on this view, are constituted by the selective success of ancestral traits. Thus the selected function of the heart is to pump blood just in case ancestral hearts pumped blood in a manner that was selectively successful. This view—the theory of *selected functions*—is compelling in two quite general ways. The first is that the biological realm has evolved and it is plausible to hold that natural selection has been an important, even if not exclusive, cause of such evolution. Of course, selection requires variation among traits with respect to their causal powers, and selection sorts among traits according to their causal efficacy. In consequence, traits that better contribute to reproductive success tend to be preserved. It thus is plausible to understand functions as those effects that enable a trait to work better than other variants of that trait. Hence the intuition that selection is a natural process capable of producing genuine functional categories defined in terms of norms of performance—an intuition defended in Millikan (1984, 1993) and Neander (1991, 1995). The second compelling feature is that function attributions are important in a wide range of subdisciplines within biology. Insofar as these subdisciplines appeal to the theory of evolution by natural selection for theoretical unity, the theory of selected functions contributes to (or at least draws on) that unification. And, of course, the theory of evolution by natural selection does indeed offer unity across a wide range of subdisciplines in biology, as Wimsatt (1997) emphasizes. So the theory of selected functions, grounded in the theory of evolution by natural selection, appears appropriate and promising.

Interestingly, the historical approach to functions, although typically cast in terms of selection, does not require it. Selection is one form of historical success, but there are others. Buller (1998) suggests that ancestral contributions to ancestral fitness suffice for the emergence of natural norms even when there is no variation upon which selection can act. This view, currently a minority, lacks the prima facie plausibility of the theory of selected functions because it leaves out the sorting powers of selection. But it has its attractions nonetheless. For, on this view, functions emerge out of a natural process that resulted in a different kind of historical success. They arise out of ancestral contributions to fitness whether or not they were favored by selection. So long as the relevant traits were heritable, their contributions to fitness did indeed contribute to their own preservation. I locate this alternative in section III under the rubric of the combination approach, since it combines the systemic approach with the historical. The important point here is that the historical approach, while typically cast in terms of selection, can be cast in terms of any natural process that produces historical success.

I begin with the theory of selected functions.[1] While this theory is embraced widely, the focus of my discussion is its status among philosophers, especially philosophers of biology and of mind.[2] The core idea of the theory is that the function of any type of trait is to do whatever ancestral tokens did that resulted in the selective success of that type. Where "trait T" and "organism O" refer to lineages, the selected function of trait T in organism O in selective environment E is to perform task F only if

(i) Ancestral tokens of T in O performed F in E,
(ii) T was heritable,
(iii) Ancestral performances of F enabled organisms with T to satisfy demands within E better than organisms without T,
(iv) Superior satisfaction of selective demands enabled organisms with T to out-reproduce (in the long run) those without T,
(v) Superior reproduction caused organisms with T and hence tokens of T to persist or proliferate in the population.[3]

This way of formulating the theory is parochial, and deliberately so. Evolution by natural selection occurs within populations of organisms, to be sure, but it occurs elsewhere as well. It can occur within populations of cell lineages that compete for resources during ontogenesis (Buss 1987); it can occur within populations of groups of organisms that likewise compete against one another (Brandon 1990; Sober and Wilson 1998). However, the attribution of functions is most common and perhaps most compelling for traits such as eyes and hearts and livers. That is why I cast the theory in terms of selection at the level of organisms, since organismic selection is caused by differences between traits of organisms. But my

1. Neander (1991) calls this the theory of "selected effects" functions and Millikan (1984) the theory of "proper" functions. I prefer Neander's more descriptive terminology. I have abbreviated it to "selected" functions.

2. In addition to Neander and Millikan, advocates of selected functions include Allen and Bekoff (1995), Brandon (1981, 1990), Godfrey-Smith (1993, 1994), Griffiths (1992, 1993), Matthen (1988), Mitchell (1995), Price (1995), Shapiro (1998), as well as Lycan (1987, 1988), McGinn (1989), Post (1991), and others. This theory is widely thought to be key to understanding the norms of nature.

3. A sixth condition is added presently.

formulation of the theory, as well as the claims I defend, can be generalized to cover selection at any level of biological organization.

Conditions (i)–(iv) incorporate the central conditions required for evolution by natural selection. Condition (iv) requires differential reproduction within the population and (iii) requires that this difference be a consequence of differences between organisms in the success with which they negotiate the demands of the selective regime. This latter difference, moreover, must be a consequence of differences in the causal efficacy of heritable traits that vary in the population. This accounts for the explicit contrast in (iii) between organisms endowed with T and organisms not so endowed. Advocates of selected functions are committed to conditions (i)–(iv) or to equivalent conditions.

The case of the peppered moth *Biston betularia* is illustrative. During the latter half of the nineteenth century in Manchester, England, there was a dramatic increase in moths with dark-colored wings and a proportional decrease in moths with light-colored wings. The cause was largely selection. As industrial coal soot darkened the tree trunks upon which the moths roosted, predatory birds increasingly spotted and consumed moths with light-colored wings. As dark coloration answered successfully the demand for camouflage, light coloration failed.[4] Dark coloration thus satisfied conditions (i)–(v): It provided camouflage, was heritable, enabled dark-colored moths to better avoid predatory birds than light-colored moths, and thus enabled dark-colored moths to outreproduce light-colored ones. This differential in reproduction caused the proliferation of dark coloration such that, by the close of the nineteenth century, 99 percent of the moths in the Manchester population had dark-colored wings. Dark coloration thus acquired the selected function of providing camouflage.

The case of the peppered moth involves a slight complication. There is a recurrent mutation within the moth population for dark coloration and this makes it unlikely that all of the dark moths in the population today are descended from a single lineage. Instead, the population comprises several sublineages of dark moths derived from several distinct mu-

4. See Kettlewell (1973) for details. For recent reflections on peppered moths, see Rudge (1999) and Hagen (1999).

tation events. The theory of selected functions, however, applies to traits related by genealogical descent, and this might be thought to raise a problem in the case of the peppered moth. But the problem is merely apparent. The several functional lineages arise from distinct mutation events and thus are historically unique, but each has perpetuated itself by virtue of providing camouflage within the same selective regime. Moreover, the propensity for this mutation is a heritable feature within the larger population. The lineages that arise from these mutations are sublineages within a larger lineage. The theory of selected functions thus must be applied to each new sublineage as it arises, though each time we will discover the same functional task.

The thrust of all five conditions can be cast in terms of a pair of distinctions from Sober (1984). Traits that satisfy (i)–(iv) produced effects that were *selected for*. Properties selected for are organismic effects that were selectively efficacious, that answered successfully the demands of the selective regime, and thereby propelled themselves and the organism into their evolutionary trajectory. They are self-perpetuating properties. By contrast, traits *selected against* produced effects that were selectively inefficacious, that answered unsuccessfully the demands of the selective regime. In the moth case, wing coloration was the target of selection, where dark coloration was selected for and light coloration selected against. Properties selected for also contrast with properties or objects *selected of*. While the former are the causes and the producers of their own selective futures, properties or objects selected of are the fortunate by-products of such selective activities. Properties or objects selected of are organismic effects that do not engage the demands of the selective regime, but nevertheless are preserved by selection insofar as they are connected by gene linkage, pleiotropy, or epigenesis to properties selected for. Properties selected of thus are dragged into the organism's selective future by association with traits that propel themselves into that future. The distinction between selection for and selection of describes two routes by which organismic traits may be preserved by selection: Those selected for are preserved by dint of their own causal powers, while those selected of are preserved because of the efficacy of the company they keep.

While (i)–(iv) describe properties selected for, (v) asserts the existence of a causal-historical relation between ancestral traits selected for and descendents of those successful ancestors. This relation is typically characterized in explanatory terms. The claim is that the selective efficacy of traits described in (i)–(iv) constitutes the explanans and the persistence or proliferation of traits mentioned in (v) constitutes the explanandum. The attribution of a selected function, on this construal, is equivalent to a selective explanation of the persistence or proliferation of the specified trait. I take it, however, that this explanatory relation is parasitic on the relevant ontological relations. Conditions (i)–(v) are explanatory only because the selective efficacy of the effects cited in (i)–(iv) caused, via the mechanisms of selection, the persistence or proliferation of traits cited in (v). It is because ancestral instances were selectively efficacious that latter-day instances persisted or proliferated, and this causal relation warrants the explanatory relation between (i)–(iv) and (v).

There is, however, a further theoretical condition required for selected functions. Advocates of selected functions assert that latter-day tokens of trait T possess the relevant selected function even when, due to various forms of incapacitation, they are causally unable to fulfill that function. They assert that selected malfunctions are possible. The justification for this claim is simple and prima facie persuasive: The causal relation between conditions (i)–(iv) and condition (v) is entirely historical. All that matters is that present tokens of T are descended from selectively successful ancestors. In particular, current capacities are irrelevant. Whether or not a token of T can perform F is quite beside the point so long as the token is a descendent of a selectively successful lineage. After all, present tokens of T do exist, at least in part, as a result of the selective efficacy of their ancestors and this is true no matter what the capacities of these tokens. We thus must add to (i)–(v) the following condition:

(vi) Superior reproduction among organisms with T, by causing the persistence or proliferation of tokens of T, imposes an office or role upon descendents of T.

The relevant role is the performance of task F. This role includes its own standard or norm of evaluation; tokens endowed with this role are "supposed to" do F. Hence the oft-repeated claim that selected functions are normative properties; they are normative to the extent they specify a norm of performance and apply that norm to descendent tokens of the functional type. The efficacy of descendent tokens is evaluated in terms of the success with which they fulfill the norms definitive of their type. And selected malfunctions are possible insofar as descendent tokens possess these sorts of normative functional roles. Conditions (i)–(vi), taken together, are necessary and sufficient for the occurrence of selected functions.[5]

The theory of selected functions is thus committed to an ontology of functional offices or roles created by ancestral selective success and imposed upon descendent tokens. Exactly what sorts of properties are involved is a topic pursued in chapter 5. Suffice it to say that, on my view, the ontology of selected functions should make us doubt the naturalistic credentials of the theory. Besides the ontology of the theory, however, two additional features deserve mention. The first is that the theory is thought to stand as an autonomous and relatively fundamental account of functional properties. Millikan, Neander, and Wimsatt, as we have seen, all claim that the theory of selected functions is more basic than its nonselectionist alternatives. Selection, it is claimed, produces functions more robust and entrenched than any other natural process. Hence the autonomy, and perhaps the primacy, of selected functions. The second feature is that the theory is thought to naturalize our concept of functions with great success. The theory explicates functions in terms of the theory of evolution by natural selection, and because the latter is a highly confirmed and well-developed theory, the former inherits its naturalistic credentials. The postulation of selected functions is thought to commit us to the same causal-historical properties as the theory of evolution by natural selection. As we shall see, however, beginning in chapter 3, neither of these purported virtues belongs to the theory of selected functions. Selected functions are not autonomous but merely one kind of systemic function, and even if selected functions were autonomous, the naturalistic status of the theory is objectionably weak, especially when compared to the theory of systemic functions.

II The Systemic Capacity Approach

Advocates of selected functions typically characterize their theory in explanatory terms. Following Wright (1973, 1976), they assert that the

5. See, however, chapter 7 for a further qualification of the theory intended to block the claim that vestigial traits can have selected functions.

attribution of a selected function is equivalent to an explanation of why tokens of the functional trait persisted or proliferated in the population.[6] As suggested above, however, the explanatory virtues of the theory are parasitic on the relevant causal powers of selectively successful traits; that is why I cast conditions (i)–(vi) in causal rather than explanatory terms. At any rate, like the theory of selected functions, the theory of systemic functions typically is characterized in explanatory terms—though in this case the characterization is more apt. Now, in chapter 4, I argue that it is a mistake to cast the theory of systemic functions in exclusively explanatory terms; the theory must include constraints on the kinds of systems to which it is rightfully applied. Nevertheless, an appeal to explanatory considerations is far more important for systemic functions than for selected functions.

The attribution of systemic functions, as usually understood, explains not the presence of any item, but rather some capacity of a system. The emphasis is on explaining how various parts of the system work together and give rise to a more general systemic capacity. Consider a cardiologist studying the circulatory system. She wants to explain how the system exercises its capacity to circulate nutrients. She analyzes the system into its various components and identifies various causal or structural capacities. If substantial background theory is available, specifying components may be a matter of course, but if the system is poorly understood, such specifications may be highly tentative. The cardiologist then traces the interactions among these various effects with an eye to identifying interactions that play a role in the exercise of the systemic capacity. This may involve various forms of inhibitory or excitatory studies of the system.[7] In the former, we inhibit certain components and examine the effects of such deficits on the exercise of the systemic capacity; in the latter, we stimulate some component and observe the effects on the systemic capacity. Studies of both sorts may warrant tentative inferences concerning the structural or causal roles of the isolated components in producing the systemic capacity. When, for example, the cardiologist comes to the heart, she discovers that the heart's pumping, unlike its thumping or vi-

6. The relation between Wright's theory of functions and the theory of selected functions is discussed in chapter 5.

7. I take these terms from Bechtel and Richardson (1993), 19--20.

brating, contributes crucially to the circulation of nutrients. Circulating blood is the main work of the heart that contributes to one of the main accomplishments of the circulatory system, namely, the delivery of nutrients. She thus attributes to the heart the systemic function of pumping blood and she can accomplish this while knowing nothing at all of the selective history of hearts.[8]

In general terms, we apply the theory by specifying the system S and the capacity C we wish to explain. We ask, By virtue of what does S exercise C? We thus might ask, By virtue of what does the human animal remain alive? or, By virtue of what does it maintain the integrity of its internal economy? We analyze the animal into component tasks—respiration, circulation, digestion, etc.—and we individuate the subsystems responsible for these tasks. We thus come upon the body's major subsystems to which we can attribute systemic functions. The circulatory system, for example, contributes to internal integrity by circulating nutrients. And now we apply our general strategy again, this time to the subsystems individuated. We analyze, for example, the circulatory system as the cardiologist would, identifying the main component tasks—assimilation of nutrients into the blood, propulsion of blood, cellular absorption of nutrients, etc.—and individuating the sub-subsystems responsible for these tasks. It is at this level that we come upon the causal effects of the heart and, on the basis of this analysis, attribute to the heart the systemic function of pumping blood. This function is attributed relative to the circulatory system's function of circulating nutrients. We may, if we wish, iterate our strategy further, analyzing various components of the circulatory system into their main subtasks and subcomponents.

The attribution of systemic functions is thus relative to the type of system under investigation, the systemic capacity we wish to explain, and the specific analysis of the system into component parts. The same trait in a different type of system may have a different systemic function or none at all. And the same trait in the same system under a different analysis—relative to a different systemic capacity—likewise may have a different systemic function or none at all. This is just to say that any one trait can and, in complex systems, often does exercise more than one systemic capacity. The attribution of systemic functions is also relative to the

8. I discuss the various theoretical roles of systemic functions in chapter 6.

appropriate level within the layered analysis. Complex systems require multilayered analyses while simple systems may require only two such layers.

A further feature of the theory of systemic functions is the importance of providing evidence for the existence of physical mechanisms within the system capable of instantiating the systemic functions attributed at the various levels in our analysis. This is an aspect of Cummins's original formulation often overlooked. Indeed, Cummins's formulation is usually described as advocating a strategy of inquiry that is merely top-down, that does little more than specify lower-level capacities in order to account for the exercise of higher-level capacities, while ignoring the mechanisms responsible for implementing these capacities.[9] But Cummins (1983) holds that any top-down taxonomy of systemic functions must be confirmed by the identification of mechanisms within the system that instantiate the functions attributed, for otherwise we have no grounds for the claim that the system in fact possesses such functions:

> Functional analysis [or systemic function analysis] of a capacity C of a system S must eventually terminate in dispositions whose instantiations are explicable via analysis of S. Failing this, we have no reason to suppose we have analyzed C as it is instantiated in S. If S and/or its components do not have the analyzing capacities, then the analysis cannot help us to explain the instantiation of C in S. This point is easy to lose sight of in practice because functional analysis is often difficult. Having, finally, arrived at an analysis of C, we are under pressure to suppose the analyzing capacities must be instantiated in S somehow. (Cummins 1983, 31)

The danger in analyzing systems from the top down is that we can construct more than one taxonomy that seems to account for the higher-level capacity we wish to understand. What we also want to know, however, is which taxonomy is correct; we want to know which account accurately represents the layered systemic capacities of the system. After all, we are after more than a possible account of the workings of the system—we want an accurate account so that we may predict and control its behavior. Hence the requirement that we adduce evidence at the level of physical

9. For example, Bechtel and Richardson refer to Cummins's strategy as "synthetic" and as "top-down" (Bechtel and Richardson 1993, 18). And, to be sure, the top-down feature of Cummins's formulation is given center stage in Cummins (1975) and (1983).

mechanisms for the functions we attribute. This is what Cummins means when he says that systemic functional analyses must terminate in dispositions the instantiation of which is explicable via an analysis of S. The analysis of S must include a specification of physical mechanisms. The analysis must include evidence concerning mechanisms that make plausible the claim that the various systemic functions in our taxonomy are in fact instantiated in S.

Thus, where "A" refers to the analysis of system S into its systemic tasks and components, and where "C" refers to the systemic capacity we wish to explicate, item I within system S has systemic capacity function F if and only if:

(i) I is capable of doing F,

(ii) A appropriately and adequately accounts for S's capacity to C,

(iii) A accounts for S's capacity to C, in part, by appealing to the capacity of I to F,

(iv) A specifies the physical mechanisms in S that instantiate the systemic capacities itemized.[10]

Iterations of this general strategy continue until they cease to contribute to our investigatory aims. At some point our analyses identify sub-sub-. . . systems that are so simple as to be uninformative vis-à-vis our explanatory goals. At this point, if not sooner, we turn our attention to locating the physical mechanisms that instantiate the functions in our taxonomy.

There are at least two importantly different sorts of systemic capacities that can qualify as systemic functions. Cummins (1983)—following Haugeland (1978)—distinguishes between "systematic" and "morphological" capacities; I will refer to them, respectively, as *interactive* and *structural* systemic capacities.[11] Interactive capacities contribute to a higher-level capacity by virtue of their causal interactions with one

10. Conditions (i)–(iii) are adopted more or less directly from Cummins (1975). Condition (iv) is added on the basis of the above excerpt from Cummins (1983). Amundson and Lauder (1994), while developing Cummins's theory in ways that are highly compelling, do not emphasize the importance of bottom-up constraints. They make no mention of the physical mechanisms required above in condition (iv).

11. Cummins (1983), 28–34.

another. The exercise of such capacities requires coordinated interactions between two or more items within the system. Thus, the pumping of the heart involves the coordinated effects of several distinguishable capacities: electrical signals sent from the brain, the expansion and contraction of muscle, the propulsion of blood, the expansion and contraction of arteries, and so on. Structural capacities, by contrast, rely less upon coordinated interactions and more upon the structure of the items involved. Haugeland (1978) offers a nice example. A fiber-optic cable has the capacity to transmit images; each fiber within the cable has the capacity to transmit light; the organized effect of all the fibers gives rise to the capacity of the cable to transmit an image. The fibers instantiate the higher-level capacity of the cable not by virtue of their interactions, but simply because the outputs of each fiber are organized in the right way. The higher-level capacity exists insofar as the structures at the lower level—the individual fibers—are organized aright. Organization of structure, rather than organization of interactions, gives rise to the higher-level capacity. This distinction between interactive and structural systemic capacities will be important in subsequent chapters.

The theory of systemic functions thus appears to contrast sharply with the theory of selected functions in at least three ways. (a) The theories appear to have distinct explanatory aims. While selected functions explain the persistence or proliferation of a trait in the population, systemic functions explain how a system exercises some capacity. Systemic functions do not explain why an item has persisted—or so it seems, but see chapter 3—and selected functions do not explain how the lower-level capacities of some system conspire to produce some higher-level capacity. (b) The theories also differ in their accounts of the origins of functional properties. The theory of selected functions asserts that selected functions are offices or roles produced by the natural selective success of ancestral tokens of the trait. The theory of systemic functions, by contrast, makes no specific requirements on the history of the functional trait. It asserts that systemic functions are specific systemic capacities—interactive or structural—within a given system, whether or not the system has been affected by selection. (c) Finally, the theories differ over the attribution of malfunctions. The attribution of a selected function involves the attribution of a norm of performance that persists when the requisite physical capacity is lost. This is thought to underwrite the possibility of malfunctions. The attribution of a systemic function, by contrast, requires satisfaction of condition (i), namely, that item I be capable of performing task F. This requirement is usually taken to mean that if the item has lost the requisite physical capacities it has thereby lost the associated systemic function, in which case malfunction cannot occur. These three differences, as well as others, tell in favor of the theory of systemic functions, as subsequent chapters endeavor to show.

Much in the systemic capacity approach is fundamentally sound. Nevertheless, the theory has come under criticism for being promiscuous in the functions it attributes and for its inability to account for the possibility of malfunctions. I shall argue, however, that the theory can be developed in ways that diminish the force of both objections. The charge of promiscuity can be blocked by making explicit the kinds of systems to which the theory properly applies. And as concerns malfunctions, we should agree that the theory cannot account for the occurrence of systemic malfunctions, but we also should insist that this is a virtue, not a vice. The attribution of malfunctions to natural, nonengineered traits is, on my view, at odds with our naturalistic approach to inquiry generally and with the theory of evolution by natural selection specifically. These developments, and others besides, result in a version of the theory of systemic functions free from the defects with which it usually is charged and free from the flaws that plague the theory of selected functions. I begin my defense of this view in chapter 3.

III Combination Approaches

The theory of selected functions and the theory of systemic functions appear suited to distinct theoretical ends. The attribution of selected functions explains why things have persisted or proliferated and why they can malfunction, while the attribution of systemic functions explains how systems exercise various capacities in terms of other, more specific systemic capacities. This has prompted several theorists to try to combine or in some way preserve the best of both approaches. These attempts fall into three rough categories. The *unification* view aims to subsume selected and systemic functions under a more general theory of functions. The *pluralist* view asserts that both theories should be preserved and employed for their distinctive investigatory ends without being unified. And the *instantiation* view asserts that selected functions are a specific instance of systemic functions and should be understood as such. I shall describe each alternative and indicate briefly where my view differs.

Unification

Despite the evident differences between the theories, Kitcher (1993) argues that we should marry selected and systemic functions under a generic concept of design: "This unity is founded on the notion that the function of an entity S is *what S is designed to do*" (Kitcher 1993, 379). The design in artifacts originates in the intentions of the designers, producers, or consumers; the design in natural objects originates in the processes involved in evolution by natural selection. If there is a suitably generic concept that subsumes design in artifacts and in natural selection—and Kitcher argues that there is—we can develop a general theory of functions according to which all functions are explicated in terms of such design. Imagine making a machine the function of which is fixed by our intentions. A screw falls accidentally into the workings of the machine during assembly. Suppose that, had it not fallen into that very place, the machine would have failed to operate as intended because some connection would have failed to obtain and, unbeknownst to us, this connection is necessary to the machine's operations. Kitcher asserts that, despite its unintended role within the machine, the screw nevertheless has a function. The specific role played by the screw may not be intended, but the function of the machine as a whole is intended, and since the screw contributes to the fulfillment of the machine's function, it thereby possesses a derived but unintended function. There can be, in consequence, unintended features of artifacts that are functional nonetheless. Hence there exists a concept of design applicable to artifacts that does not appeal directly to the effects of a designer. And, according to Kitcher, an analogously derived concept of functions—one that appeals to the designing effects of natural selection rather than the intentions of designing agents—applies to a host of natural, nonartifactual objects.

This push toward unification is driven by Kitcher's (1981) view that a scientific theory increases in explanatory power as it unifies an increasingly large range of phenomena under an increasingly narrow range of laws or principles. While I am sympathetic to unification as a constraint on some forms of scientific explanation, I am opposed to this particular application, for I do not believe that we can unify selected and systemic functions under a generic concept of design. I defend this claim in the next chapter. More generally, and as I argue throughout subsequent chapters, the theory of selected functions boasts of various virtues that it fails to deliver. It is a failed theory. And to the extent that Kitcher's proposed unification rests substantively on a failed theory, then it too fails. This is a defect afflicting the combination approach generally.

Pluralism

Pluralism is popular these days in settings beyond the political. It is popular in accounts of species, the units of selection, and the nature of functions. Godfrey-Smith (1993) and Preston (1998) argue for a pluralist view of functions—one that holds selected and systemic functions in a single embrace—and Amundson and Lauder (1994), along with a reformed Brandon (forthcoming), express approval for a similar sort of pluralism. The pluralist asserts that both theories of functions are required because each applies to distinct phenomena. Godfrey-Smith appeals to the distinction marked by Mayr (1961) and Tinbergen (1963) between *evolutionary explanations* and *functional explanations* in evolutionary biology and behavioral ecology. The former explain in historical terms how organisms came to have the features they have, while the latter explain how a given trait exercises certain of its capacities. The differences between selected and systemic functions map neatly onto the difference between evolutionary and functional explanations. The attribution of selected functions is equivalent to an explanation of why the functional trait persisted or proliferated and the attribution of systemic functions explains how certain systems exercise certain of their capacities. The pluralist's thesis, then, is that both theories of functions are required in order to preserve a fruitful distinction employed among practicing biologists. Moreover, the sort unification advocated by Kitcher is rejected on the grounds that it conceals, rather than uncovers, important causal differences between such phenomena.

Preston's general argumentative strategy is similar to Godfrey-Smith's, though she makes no appeal to the distinction marked by Mayr and Tinbergen. She argues instead that the theory of selected functions as developed by Millikan (1984, 1993) is internally inconsistent as a consequence of attempting to extend its reach to traits that lack the relevant sort of selective history. Some functions, despite Millikan's best efforts, cannot be captured within the theory of selected functions. The mantling behavior of some herons, for example, enables them to shade the water and thus better spot fish. The systemic function of such behavior is evident. Preston claims, however, that it is not plausible that this behavior was selected for its utility in spotting fish, since related species lack this behavioral trait.[12] These and other considerations are intended to convince us that each theory of functions is best suited to distinct sorts of phenomena and thus that both should be preserved intact.

I agree that unifying the theories under a generic concept of design conceals causal differences of theoretical importance. Nevertheless, I disagree with the pluralist's position. My view is that the theory of selected functions is a failed theory; it simply does not have the virtues that advocates claim for it and thus is not worth preserving. In consequence, any attempt to couple selected functions with any other type of functions is doomed at the outset, no matter how plausible the additional theory. We do better to dispense with selected functions altogether and focus our efforts on developing the theory of systemic functions. This is especially compelling in light of the claim, defended in the next chapter, that the theory of systemic functions renders the theory of selected functions explanatorily redundant.

12. Preston writes: "While mantling is obviously a useful thing to be able to do, the fact that only one species of heron does do it indicates that the selection pressures favoring it are not very strong. Since wings of birds that do not fly rapidly become vestigial, the pressure would have to be strong to maintain wings suitable for mantling in a flightless heron. So the wings of (some) herons are an ongoing system-functional exaptation for mantling" (Preston 1998, 240–41)—which is to say that mantling has a systemic function but not a selected function. See Millikan (1999) for her response to Preston's criticisms.

Instantiation

A few recent views assert that selected functions are a specific instance of systemic functions. These views are near my own. Indeed, I argue in chapter 3 that the theory of systemic functions, being more general in scope than the theory of selected functions, applies to populations and their capacity to evolve or remain in equilibrium as a consequence of natural selection. And this, I claim, renders the theory of selected functions explanatorily redundant and thus eliminable. The redundancy of selected functions is especially important in light of the thesis, defended in chapter 7, that the theory of selected functions lacks the internal resources with which to account for the possibility of malfunctions. Core elements of the theory, I claim, are simply beyond repair. The only aspect of the theory worth preserving is the idea that some functions arise by virtue of the effects of selection on a population, and this thought is captured rather neatly in the theory of systemic functions. It is in this sense that selected functions, or some remnant of selected functions, may be understood as instantiating the theory of systemic functions. Before turning to that argument, however, I want to sketch several versions of the instantiation view and indicate how they differ from the view developed here.

Griffiths (1993), to the best of my knowledge, was the first to suggest explicitly that selected functions are a specific type of systemic functions—though see Enç (1979). Griffiths allows that systemic functions apply quite broadly, to systems affected by selection and those not so affected. Systems affected by selection include organisms. The relevant systemic capacity we wish to explain is relative survival and reproduction. We analyze relative fitness into various fitness components, namely, the effects of traits that contribute to fitness. Such effects qualify as systemic functions. Griffiths thus holds that selected functions are a specific kind of systemic function. He also holds that we can, from the perspective of systemic functions, preserve the apparent normativity of selected functions and extend the theory of selected functions to the functions of artifacts.

Walsh and Ariew (1996) propose a view similar to Griffiths's. The relevant system is the organism and the relevant systemic capacity is differential reproduction. As with Griffiths, we analyze the organism to identify lower-level capacities that contribute to the relevant rate of reproduction. The systemic capacities identified are selected systemic functions. And like Griffiths, Walsh and Ariew hold that the resulting view contains the resources with which to account for the possibility of malfunctions. Interestingly, and despite the appeal to selection, this view does not fit neatly the historical approach described in section I. Selected systemic functions, on this view, are the products of selection but not necessarily selection in the past. If tokens of trait T are currently being selected for task F, then, even if they have never been so selected before, T now has the selected function of performing F. Selection is essential, but selection in the past is not. By contrast, Buller (1998) holds that the past is essential, but selection is not. He distinguishes between what he calls the "strong etiological theory"—the view that selection is necessary for functions—and the "weak etiological theory"—the view that some sort of historical success, but not necessarily selection, is essential for functions. Buller endorses the latter. As with Griffiths, an organism is subject to a systemic analysis. This time, however, the systemic capacity we wish to explain is not differential reproduction, but organismic fitness. A trait can contribute to fitness even when no selective pressures affect it. It can contribute to fitness by contributing, say, to viability or fecundity, and it can accomplish this even when it exists in the population in only one form. It thus can facilitate its own etiology. Weak etiological functions are those systemic functions that contributed to the fitness of heritable ancestral traits. And like Griffiths, Buller holds that malfunctions are possible.[13]

I agree with these versions of the instantiation view that the theory of systemic functions is far more general than the theory of selected functions and, in fact, subsumes it. But I do not agree that the theory of selected functions, taken in its entirety, or simply the core elements of selected functions, such as their alleged normative status, ought to be

13. How can a trait be heritable if it exists in the population in only one form, since heritability, by definition, requires variation? Buller suggests two possibilities. It may happen that the necessary mutations required to produce an alternative to trait T, while genetically possible, never in fact occur. Or it may happen that the mutations occur but T and the alternative to T never coexist in the same selective environment. In either case, we have genetic variability that makes no selective difference. See section 2 of Buller's discussion for details.

preserved within the theory of systemic functions. These versions of the instantiation view endeavor to preserve central aspects of the theory of selected functions. They especially wish to preserve the intuition that some functional properties are "proper" and "teleological," endowed with normative standing, while others are useful but not genuinely purposive. The thought is that all systemic functions are useful, but only one type of systemic function—those that arise by being selected for—are genuinely purposive. On my view, however, these versions of the instantiation approach fail to appreciate the substantial defects in the theory of selected functions. As I argue in later chapters, selected functions fail to fit comfortably a naturalistic approach to inquiry, especially when compared to systemic functions; they fail to account for the possibility of malfunctions; and they are theoretically redundant on systemic functions. These versions of the instantiation view ought to be discarded on the same grounds that selected functions should.

It is worth mentioning a fourth view that falls somewhere between the historical and combination approaches. As developed by Bigelow and Pargetter (1987), the "selective propensity theory" appeals to selection but makes no essential appeal to past history. Unlike the historical approach, the selective propensity view does not require actual selective success, only the propensity for such success. A trait has a function, according to Bigelow and Pargetter, so long as it has the propensity to be selected for the relevant effect in its natural habitat. If the habitat is normal and if accident does not befall the organism, then such propensities will be selected for; but if something thwarts the selective potential of the relevant trait, that does not diminish its functional status, since it would have been selected for under the appropriate circumstances. Functions thus are subjunctive properties; they exist whether or not ever manifested. The actual selective future of a trait type is not relevant. Potential is everything.

It is tempting to place this view in the instantiation category because, like the theory of systemic functions, it appeals to the propensities of traits to exercise certain capacities within the organism which in turn contribute to the higher-level capacity of reproduction. The similarity of Griffiths's view to Bigelow and Pargetter's is clear. Bigelow and Pargetter do not explicitly exploit this similarity to the theory of systemic functions and they may, if pressed, wish to reject it. Nevertheless, they assert that

> functions can be ascribed to components of an organism, in a descending hierarchy of complexity. We can select a subsystem of the organism, and we can ascribe a function to it when it enhances the chances of survival in the creature's natural habitat. Within this subsystem, there may be a subsubsystem. And this may be said to serve a function if it contributes to the functioning of the system that contains it . . . and so on. (Bigelow and Pargetter 1987, 193)

This is to take a type of organism as the relevant system and organismic survival as the relevant systemic capacity, and then analyze the system into subsystems in order to identify salient systemic contributions vis-à-vis survival. So selective propensity functions, like systemic functions, involve the current existence of the relevant propensity.

At the same time, however, this theory does not fit comfortably the combination approach because of the importance it assigns selection. Selective propensity functions are dispositions of organisms, to be sure, but they are functional only if they have the propensity to be selected for under normal environmental conditions. This is to assert that the only natural process capable of producing functional properties is selection. That is an excessively narrow view of the nature of functions. On my view, systemic functions are broader in scope than selected functions, for they can emerge in systems not affected by selection. Functions emerge in the operations of hierarchical systems; they emerge in the context of various lower-level capacities performing those tasks that give rise to higher-level systemic capacities. And they emerge whether or not selection is affecting the system. So, like the theory of selected functions, Bigelow and Pargetter's fixation on selection results in a theory the scope of which is unacceptably narrow.

I mention the selective propensity view, however, not merely because it strains the categories into which I am trying to fit the many extant theories of functions, but because there is, in my view, something right about it. As Bigelow and Pargetter emphasize, selective propensity functions are forward-looking. They are dispositions the effects of which have a promising future, and if the organism's habitat does not change significantly, this promise will be actualized. Now I do not fully endorse this particular way of explicating the forward-looking aspect of functions, but I do appreciate the importance of the general insight that functions include some sort of reference to or expectation concerning the future. I offer my own take on this insight in chapter 6.

The final version of the instantiation view—the "systems-theoretical" view—has been developed by various authors, most recently by Schlosser (1998). Previous advocates include Nagel (1977) and Enç (1979). On this view, functions are properly attributed to states or to traits that "re-produce" themselves in the context of systems that are sufficiently complex. Self-re-producing states cause themselves to recur within a system. The beating of a token heart contributes to the maintenance of life, which causes more beats of that token heart. Self-re-producing traits cause themselves to recur across generations by virtue of their selective efficacy. The beating of the heart contributes to survival and to mating, which contributes to the production of descendent hearts and thus descendent heartbeats. "Re-production" thus applies to repeatable states within a system and to the recurrence of a trait throughout a lineage. A system is sufficiently complex so long as the relevant state or trait is re-producible in more than one way depending upon environmental conditions. Our thermoregulatory system maintains an internal body temperature one way in heat (perspiring) and another in cold (shivering), in which case it is complex in Schlosser's sense ("plastic" in Nagel's sense). This notion of complexity admittedly is not precise and thus there are likely to be cases in which the attribution of a function is hard to decide.

This view applies generally to complex systems with self-reproducing states or traits, including the reproduction of organisms and organismic traits across generations. In consequence, selected or weak etiological functions can be seen as a kind of systems-theoretical function. The selected function of the heart is to pump blood if pumping blood causes, via selection, the reproduction of descendent hearts. The weak etiological function of the heart is to pump blood if pumping blood causes, via contributions to fitness, the reproduction of descendent hearts. In either case, we identify the way in which the heart contributes to the reproduction of hearts—by selection or by contributing to fitness. The absorption of historical functions into the systems-theoretical approach thus parallels other versions of the instantiation approach.

On my view, however, the systems-theoretical view—like the view in Enç (1979)—is subject to an objection that does not affect the theory of systemic capacity functions. The objection is that the process of reproduction is not necessary for the existence of systemic functions. I grant that traits with systemic functions often do contribute to their own recurrence in the ways described by Schlosser. This is especially the case for the traits of living things, since they tend to reproduce themselves anyway. But the ubiquity of reproduction among traits with systemic functions is incidental, not essential, to the possession of those functions. Possession of systemic functions requires contributions to the exercise of some higher-level systemic capacity; this is true even when the functional trait fails to contribute to its own recurrence. After all, there are capacities of systems we wish to explain other than the capacity to persist over time, in which case the recurrence of the functional trait is not always to the point. Among organisms that tend to reproduce themselves, traits that contribute to the various operations of the organism tend to be reproduced. But that, as I say, is a byproduct and not an essential part of the conditions that give rise to systemic functions. I defend this claim in chapter 4.

IV Conclusion

These, then, are the main approaches to functions extant in the literature—historical, systemic capacity, and combination. And these are the main topics of discussion in subsequent chapters. As noted, I intend to develop and defend the systemic capacity approach. I begin my defense by showing that this theory, which applies quite broadly, applies in particular to populations and their members. This application shows that selected functions, or certain remnants of selected functions, are best understood as systemic functions that emerge in systems affected by selection. Otherwise the theory of selected functions is merely redundant on the theory of systemic functions and ought to be discarded outright. I turn now to defense of this claim.

3

Discarding Selected Functions

Among the approaches to functions described in the previous chapter, the historical and the systemic capacity approaches are thought to be particularly compelling. They are also thought to be distinct from one another. Each approach has been represented as an independent, self-sufficient account of functions. This alleged distinctiveness raises the question, Are these approaches (1) in competition, (2) complementary, or (3) unified by some more-general theory? All three answers have been defended or suggested in the literature. The aim of this chapter is to defend a fourth: Contrary to appearances, there are not two distinct theories of functions on offer, only one—the theory of systemic functions. Now, in later chapters, I argue that the theory of selected functions is deficient in various ways. I argue that the theory has relatively poor naturalistic credentials and that it lacks the resources with which to account for malfunctions. There is, in consequence, rather little in the theory worth preserving. The thesis of this chapter is that the theory of selected functions, when construed as a separate autonomous theory, is a failure and hence ought to be discarded. This leaves room for functions that arise in systems affected by selection, but those are best understood as nothing more than selectively efficacious systemic functions. There are, in consequence, not two theories of functions on offer, only one, and it accounts for the full range of functional properties discoverable by life scientists.

This way of conceptualizing the relationship between the (allegedly) two theories has considerable advantages. One is that it unifies our theory of functions. The prevailing view is that we have two quite separate notions of functions operative in biology and perhaps in other areas of inquiry as well. I intend to show that this is not so. And I take it for granted that an increase in theoretical unification is, other things equal, a desirable end. Another advantage is that this view brings into focus the nature of functional properties. The unification offered does not conceal or gloss over the structure and efficacy of functions; it makes these features of functions more explicit. The increase in conceptual clarity comes by seeing that certain effects count as functional not because they are (or were, or would be) selectively (or etiologically) successful, but because they contribute to the exercise of some more-general capacity of the system in which they operate. Selective success is allotted no special status. Functions exist insofar as component capacities contribute to larger systemic capacities and this is true for systems affected by selection as well as those not so affected.

As indicated earlier, I intend to develop an alternative version of the theory of systemic functions in subsequent chapters. For purposes of this chapter, however, I employ Cummins's original formulation of the theory and defend the thesis that selected functions ought to be discarded in favor of Cummins-style systemic functions. This strategy enables me to note the objections typically pressed against Cummins's formulation and direct the reader to later chapters in which I address these charges. Moreover, the version of the theory that I develop throughout the book retains the basic structure of Cummins's formulation. So, if I succeed in showing that Cummins's account of systemic functions is preferable to the theory of selected functions, it is easy to show that my account also is preferable.

Finally, the arguments for the thesis of this chapter are involved and the considerations raised here are preliminary. Additional arguments occupy remaining chapters. Indeed, the book as a whole consists of a series of arguments in favor of systemic functions and against the historical and combination approaches. This chapter is best seen as providing a general framework within which the arguments of the remaining chapters are effectively brought to bear. I begin with an overview of the purported relationships between the (allegedly) two theories.

I The Theories and Their Relationship

The theory of selected functions asserts that the function of any trait is to do those things ancestral tokens did that resulted in relative selective efficacy. While the theory is best cast in causal-historical terms—see the formulation in chapter 2—typically it is cast in explanatory terms instead. This, I suspect, is a bit of theoretical inertia. The theory is heir to the account of etiological functions formulated by Wright (1973, 1976), according to which the attribution of a function to trait T is equivalent to an explanation of why T persists in the population. Thus, the question "What is trait T for?" is equivalent to the question "Why is trait T here?" and presumably a correct answer to one is a correct answer to the other. To attribute a function to trait T is to give an account of the reasons why T is where it is. To attribute the function of pumping blood to the mammalian heart is to assert that token hearts exist today because the pumping of blood by ancestral tokens contributed to the perpetuation of the type. Now Wright's arguments depend upon considerations of ordinary language that no longer wield much philosophical clout. Nevertheless, advocates of selected functions sympathize with Wright's view that attributions of functions are equivalent to explanations of the functional item's existence. What they add is that the relevant sort of explanation is one that appeals to the mechanisms postulated in a highly confirmed scientific theory, namely, the mechanisms of evolution via natural selection.

While the theory of selected functions typically is cast in terms of explaining the persistence or proliferation of the functional type, the theory of systemic functions aims to explain how a system exercises some higher-level capacity by appeal to the organized exercise of lower-level capacities. Advocates of selected functions explain, for example, why the mammalian heart has persisted, while advocates of systemic functions explain how the circulatory system exercises the capacity of circulating nutrients by appeal to the organized interactions among the heart, brain, blood, arteries, etc. So long as the circulatory system analyzes into components the capacities of which account for the circulation of nutrients, and so long as the heart's pumping blood is part of the explanation of that systemic capacity, the attribution of a systemic function to the heart is secure. The only other requirement is the specification of systemic mechanisms responsible for the specified capacity of the heart. The emphasis, then, is on explaining how a system works in the production of some systemic effect. And the attribution of systemic functions is always relative to an analysis of the system with respect to the systemic capacity we wish to explain.

The theory of systemic functions thus appears to contrast sharply with the theory of selected functions with respect to its explanatory aims and its explanatory strategy. To explain why an object persists is not to explain its possession or its exercise of any such higher-level capacity and to explain how an object exercises some higher-level capacity is not, it seems, to explain why it persists. So the aims of the theories seem to differ. Moreover, while the theory of selected functions explains by appeal to the mechanisms of selection, the theory of systemic functions explains by appeal to component capacities and the interactions of these capacities that give rise to the exercise of some higher-level systemic capacity. Appeals to elements of instantiation do not involve appeals to the mechanisms of selection and appeals to the mechanisms of selection do not, it seems, involve appeals to elements of instantiation. The contrast between the theories appears stark.

What, then, is the proper relationship between these theories? One option is to see them as competitors for the correct explication of our concept of functions. This assumes, perhaps illicitly, that there is a core to our concept of functions and that an adequate theory ought to capture this core. This, at any rate, seems to be a working assumption in Wright (1973), given the methods of conceptual analysis he employs. Moreover, Millikan (1989) and Neander (1991) appear to pit selected functions and systemic functions against one another, at least with respect to the attribution of malfunctions. Both assert that the theory of selected functions is preferable because it, unlike the theory of systemic functions, can account for the attribution of malfunctions. This is to assume that a proper task of a theory of functions is to preserve the intuition that functions include norms of performance that can be violated, where such norms make possible the occurrence of malfunctions. This is to construe the two theories as competing accounts of certain phenomena.

But there are alternatives to the view that these theories are competitors. As described in chapter 2, some theorists see the theories as complements to one another. Amundson and Lauder (1994), after developing Cummins's account of systemic functions in the context of anatomy and physiology, suggest that their view should supplement, not supplant, the account of selected functions defended in Millikan and Neander. Brandon (forthcoming) agrees. Godfrey-Smith (1993) claims more explicitly that each theory pursues distinct explanatory aims and, in consequence, each has the potential to provide importantly different sorts of causal information. Preston (1998) urges the same sort of pluralism on the grounds that the theory of selected functions fails to account for the full range of functional properties exemplified by living things and especially by artifacts. Kitcher (1993) disagrees with any such pluralism. There is, he claims, a more-general theory of functions that unifies both theories and, in consequence, there is but one true theory of functions, of which selected functions and systemic functions are specific instances.

These, then, are three views of the proper relationship between the theory of selected functions and the theory of systemic functions. While there are significant considerations that favor these three, there is a fourth view of the relationship that is preferable. Rather than critically engage these three, I shall explicate the alternative and a few of its most obvious virtues. This should suffice to show that this fourth view is preferable despite the considerations that favor the others.

II The Alternative

On my view, the proper relationship between the theory of selected functions and the theory of systemic functions is that there are not two theories, only one. The theory of systemic functions is the one true theory of functions. The attribution of selected functions, or at least of functions that emerge in systems affected by selection, is properly explained from within the theory of systemic functions. Anything of value in the theory of selected functions is preserved in the theory of systemic functions, while the problematic features of selected functions are discarded. The thought that some functions arise as a consequence of selection is not abandoned, but recast from within the theory of systemic functions.

The theory of systemic functions applies easily to systems that produce things. It applies to the mammalian circulatory system, which produces the regular distribution of nutrients and removal of wastes within the organism. The theory also applies to assembly lines, as Cummins (1975) points out. Producing things is the primary capacity of assembly lines and consists quite generally in the organization of component parts and the capacities of those parts. We apply the theory of systemic functions by analyzing the line's capacity into its most salient tasks and identifying components that, appropriately organized, execute these tasks. And each component, in accomplishing its specific task, exercises certain subcapacities that can be analyzed in the same way. This gives us our analysis of the assembly line and we might, as Cummins notes, employ a flow chart to represent the resulting taxonomy. Of course, not all of the capacities of any given part bear on the line's capacity to produce its products, only some do so, and only those qualify as systemic functions.

The theory of systemic functions, then, applies to systems that produce things. But one thing a system might produce is change. Another is resistance to change or a tendency toward stasis. In particular, a system might possess the capacity to produce change within or to itself, or it might have the capacity to produce an internal equilibrium that resists change. Instead of assembly lines and the products they produce, consider a population in its selective environment and the systemic changes wrought by natural selection. Consider specifically the evolutionary change wrought by certain differences in the causal capacities of varying, heritable traits within the population, or the resistance to change wrought by differences in the capacities of varying, heritable traits. The former is an instance of directional selection, the latter of stabilizing selection. Both types of selection arise from causal capacities of members of the population; both give rise to a higher-level capacity of the population as a consequence of specific components.

We apply the theory of systemic functions to a population in its selective environment the same way evolutionary biologists study change or the absence of change in a population. Suppose, for example, we want to explain the population's having evolved in some way. We ask, By virtue of what was this capacity for change exercised? To simplify, suppose the population consists of two categories of organisms classified on the basis of differences in reproductive output. Some reproduced at rate $n + 1$, others at rate n. Suppose further that ecological studies show that this difference in reproductive output was caused by differences in organismic traits, at least some of which were heritable. Those organisms that reproduced at $n + 1$ did so because of the relative efficacy of trait T; other organisms lacking T did not fare so well. These two groups, then, individuated in terms of differences in traits that resulted in differences in rates of reproduction, are the component parts that most directly account for the population's exercise of its capacity to evolve. This is the first step in applying the theory to the population.[1]

Of course, the evolutionary change we wish to explain may not be caused by direct interactions between organisms with T and those without T. One set of organisms may out-reproduce another because they are taller or smarter (or whatever) and gain better access to nutrition. This may occur without direct causal interactions between organisms. Evolutionary change thus may be caused by nothing but the sum effects of both sorts of organisms interacting within a common selective demand. And in such cases, it may be appropriate to attribute structural systemic functions rather than interactive systemic functions. Structural functions accrue to components on the basis of certain structural features rather than interactions with other components. Such functions arise because of the morphology of the components involved, not because of their interactions. From the perspective of evolutionary biology, then, it is plausible that organisms with T constitute one structural feature of the population while organisms without T constitute another. Like Haugeland's fiber-optic cable, these components may produce certain effects the sum of which gives rise to the population's capacity to evolve. We thus are warranted in attributing structural systemic functions at this first level of organization.

This is to construe a population as a kind of system composed of certain structural parts. To make this plausible, we need not formulate necessary and sufficient conditions for the existence of systems. We need only note that evolutionary biologists treat populations as systems endowed

1. Population geneticists describe evolution in terms of changes in allele frequencies and inquire into the selective pressures that caused these changes. This is consistent with my account of the structure of a population insofar as selective changes at the allelic level are mediated by selective interactions between phenotypes and the selective environment. This is so no matter what kind of items are affected by selection: There will be some "phenotypic" features, especially those directly involved in reproduction, mediating the selective efficacy of the relevant type of "replicators" (to borrow from Hull [1981], Brandon [1982], and Dawkins [1982]). I take it for granted that the mediating effects of phenotypes are implicit (if not explicit) in the strategies employed by population geneticists.

within internal structure and that they do so in ways that bear theoretical fruit. The evolutionary effects of selection are usefully modeled by dividing the population into collectives of organisms distinguished by differential reproduction. The salient structural features of these components may change as the population evolves, in which case the values employed in the relevant models will change too. As Lloyd (1988) and Thompson (1988) have argued, evolutionary biologists conceptualize populations as systems composed of various structured units, where the structural features or interactions among these units give rise to the behaviors of populations we wish to understand. This, they claim, explains why we do better to understand evolutionary theory as employing the so-called "semantic" account of scientific theories, as opposed to the "syntactic" or "received" account inherited from logical positivism.[2]

Of course, systemic features of a population may differ from systemic features of an organism. Organisms have important structural features, to be sure, but the interactions among these features are integral to animal life, while such interactions may be less central to the integrity of a population. Much depends on details of the population. A population that includes, say, highly structured social interactions may approximate or exceed the interactive complexity of some organisms. But organisms as internally complex as *Homo sapiens* probably exceed even highly structured populations in degree of interactive cohesiveness among lower-level components. At any rate, the distinction between structural and interactive systemic capacities is intended to capture the range of interactive cohesiveness among populations as well as among organisms and other more paradigmatic systems. Moreover, in the course of explicating the concept of individuality, Ghiselin (1997) describes different degrees of

2. On the received view, theories are understood as sets of sentences (or linguistic entities of some sort) related by various logical relations, where some sentences have empirical content. On the semantic view, by contrast, theories are models or sets of models that are more or less isomorphic to the phenomena being studied. The closer the isomorphism, the better the model. A model of genotype redistribution in a population over several generations is fruitful only if the structural features of the model are isomorphic to structural components of the population responsible for the evolutionary change. The semantic approach originates in the work of Suppes (1967) and is developed by van Fraassen (1972) and Suppe (1979).

cohesiveness among complex entities. He distinguishes between "cohesive individuals" (organisms) and what he calls "historical individuals" (species, lineages, populations), and portrays their differences vividly.[3] Whether or not we should conceive of species as individuals, as he suggests, is not the issue here. At issue is that his distinction between kinds of individuals provides ample theoretical room for construing populations as systems with components endowed with structural or interactive systemic functions.

Populations evolve as an effect of mechanisms other than selection. Populations change due to drift, for example. Drift produces evolution when some extraorganismic effect results in differential reproduction. Significantly, the theory of systemic functions is sufficiently general to warrant the attribution of functions in the context of evolution due to drift. By specifying the proper system—the population and relevant extraorganismic factors—we can construct an analysis of the redistribution of genotypes or phenotypes caused by drift and then attribute systemic functions to the itemized components. Such functions accrue to those components the effects of which gave rise to the specified evolutionary change. This shows that the range of systemic functions in a population is broader than the range of selected functions. More generally, the theory of systemic functions applies to systems affected by selection and to systems affected by other forces as well. It applies so long as the exercise of some higher-level systemic capacity is explicable in terms of the structural or interactive effects of lower-level capacities. Whether or not selection affects the system need have no effect on the workings of the system's various capacities. Selected functions are a specific kind of systemic function, but only one kind among several.

To return to the main thread, we now iterate our systemic strategy. We construct an analysis of each group of organisms identified at the first level. Take those organisms that reproduced at rate $n + 1$ and ask, By virtue of what did token organisms of this type reproduce at this rate? The answer is that they enjoyed superior reproduction by virtue of the causal effects of certain traits—in particular, trait T. These are traits the

3. See Ghiselin (1997), especially chapter 4. I am grateful to David Buller for directing me to this discussion.

effects of which enabled such organisms to better satisfy the relevant selective demands. Exactly which traits are involved will depend upon the details of the organism and its selective regime. In particular, traits involved will be individuated by what Brandon (1990) calls an "ecological explanation" that identifies the selective demands and organismic traits that answered successfully these demands. Moths with darker wings, for example, in the species *Biston betularia,* were better able to avoid predators by providing superior camouflage, and darkened trees trunks and predatory birds were the main ecological features that favored dark over light coloration.[4] Thus, if trait T happens to be wing coloration in this population, we have completed our analysis and can attribute to dark coloration the structural systemic function of providing camouflage. But suppose we are interested in other traits less directly involved in the organism-environment interactions. Perhaps we want to know the function of various genotypes relative to the capacity for camouflage. In that case, we iterate our strategy yet again, this time breaking into the internal economy of the organism. We ask, By virtue of what did moths with dark wings acquire their dark coloration? The answer, let us suppose, appeals to the effects of certain genes that code for dark pigmentation in concert with genes that code for wing tissue and perhaps in concert with regulator genes. We thus attribute to the specified genes the appropriate systemic functions.

This way of conceptualizing functions is at odds with the constructive suggestions of Bock and von Wahlert (1965). They suggest we pry apart two importantly different notions in biology, namely, "function" and "biological role." The difference between these concepts maps onto the distinction, marked by Mayr (1961) and Tinbergen (1963), between functional and evolutionary explanations (see section III below). They define "function" as follows:

> *In any sentence describing a feature of an organism, its function would be that class of predicates which include all physical and chemical properties arising from its form (i.e., its material composition and arrangement thereof) including all properties arising from increasing levels of organization, provided that these predicates do not mention any reference to the environment of the organism.* (Bock and von Wahlert 1965, 274)

4. See chapter 2, section I, for details. See Kettlewell (1973) for the full story.

Functions are attributed to the effects of traits affecting only the internal economy of the organism. The effects of traits extending beyond the organism to features of its environment are not among the functions of such traits, but rather among their biological roles. The latter include the roles that certain traits play in the life of the organism and, in particular, in the process of conferring a selective advantage. On this construal, functions are intraorganismic systemic capacities and carry no overtones of purposiveness or teleology, while biological roles are teleological insofar as they contribute to the persistence or proliferation of the organism. Bock and von Wahlert recommend this distinction precisely because it helps us keep separate the teleological and the nonteleological.

On my view, however, no such distinction can be sustained. Two features of my view are relevant. First, and as I argue in later chapters, I reject the assumption that nonengineered natural traits possess genuine purposes or norms of performance. I thus reject the premise that we must make room for any such teleology in our theories of natural traits. Second, and more to the point of this chapter, I reject the premise that there is a strong distinction between intraorganismic functions and extraorganismic biological roles. Leaving aside assumptions of teleology, we can explicate both properties from within the theory of systemic functions. Intraorganismic functions are systemic functions that contribute to some more general organismic capacity; the capacity of the circulatory system is illustrative. Extraorganismic roles are systemic functions that contribute to some more general capacity of the population (or some such system); the efficacy of dark coloration in the peppered moth population is illustrative. The systems being studied differ—organisms and populations—but there is no difference in the kind of function ascribed. The ascribed functions are systemic capacities that contribute to a more-general systemic capacity. We thus have no need of a strong distinction between functions and biological roles.

The basic strategy for applying the theory of systemic functions to a population affected by selection thus has three rather general steps.

(1) Analyze the population into distinct collectives on the basis of differences in reproduction, where these differences are due to differences in heritable traits that vary and give rise to evolution or equilibrium in the population as a whole.

More generally, analyze the population into whatever structural units are required to account for the capacity of the population we wish to explain. The warranted systemic functions may be structural in character.

(2) Analyze the selectively successful category (or categories) of organisms into all those components that give rise to the capacity to reproduce at the specified level.

This is to focus specifically upon systemic functions that contribute to selective success. We thus analyze organisms with T into those components instantiating its capacity to reproduce at $n + 1$. This yields a taxonomy of components to which we can attribute structural or interactive systemic functions. In the moth case, wing color has a structural systemic function, but in others—for example, selection for the capacity to process an ingested toxin—the relevant components will include interactive capacities—for example, capacities of gastric juices.

(3) If we want to know the systemic functions of traits further embedded in the organism's economy, analyze the components identified in step (2) until our taxonomy identifies the relevant lower-level traits and their salient capacities.[5]

These strategic steps illustrate a fact of considerable importance, namely, that every selected function is, at most, one instance of a kind of systemic function. Two considerations make this clear. First, the theory of systemic functions warrants the attribution of any and all functions attributed from within the theory of selected functions. Any trait affected by selection can be analyzed in terms of its systemic effects. The recursive nature of the theory's basic strategy permits us to persist in our analyses until we either lose interest or come upon components too simple to analyze further. We thus can reach any trait in which we have an interest by applying our analysis to the population, as in step (1), to its constituent structural types, as in (2), and to the organism at any or all levels of

5. These steps are deliberately abstract and tentative. The central point is that populations are treated as systems that exercise certain capacities as a consequence of the capacities of certain structural units operating at lower levels of organization. The analogy to assembly lines seems a good one.

organization, as in (3). Second, all of this is achieved within the context of explaining how the population evolved (or remained the same) via selection; all of this is achieved in the context of explaining *the very thing* that advocates of selected functions aim to explain—to wit, the effects of selection. Indeed, identifying systemic functions in the context of explaining how the population evolved (or remained the same) enables us to explain why the relevant trait persisted or proliferated in the population—just as advocates of selected functions desire. And this means that the theory of systemic functions, applied to systems affected by selection, accounts for everything that the theory of selected functions explains. And it does so entirely from within the theory of systemic functions. I conclude, therefore, that we should embrace the theory of systemic functions and dispense with the theory of selected functions construed as an autonomous theory on the grounds that there is no independent theoretical purpose served by the latter theory.

It will, of course, be objected that there is at least one thing that the theory of selected functions can account for that the theory of systemic functions cannot—the possibility of malfunctions. The consensus view is that selected functions, being the products of the process of selection, are thereby normative in character. Tokens with selected functions are "supposed to" perform their functions, since performing such functions explains why the type has persisted or proliferated. By contrast, systemic functions are thought to be fundamentally nonnormative. The theory of systemic functions applies to systems not affected by selection and thus attributes functions to traits that have no natural purposes. The objection, then, is that there is at least one theoretical purpose that only selected functions successfully address, in which case the theory ought to be retained. In fact, however, this objection is in error. As later chapters endeavor to show, very little in the theory of selected functions can be preserved. In particular, the claim that selected functions are normative—endowed with a norm of performance—cannot be sustained. In chapter 5, I argue that our naturalistic commitments conflict with the attribution of selected malfunctions and, in chapter 7, I argue that the theory of selected functions lacks the internal resources with which to warrant the attribution of any malfunctions. On my view, then, the theory of selected functions is no better than the theory of systemic functions in accounting for malfunctions. The claim that selected functions offer something that systemic functions cannot is found wanting. So the conclusion drawn above stands—the theory of selected functions, construed as an autonomous theory, ought to be discarded on grounds of redundancy.

This redundancy derives from the fact that systemic functions are more general in scope. As we have seen, populations of organisms can instantiate the capacity to evolve via differences in lower-level capacities; populations of organisms can also instantiate the capacity to maintain a certain distribution of traits via the exercise of certain lower-level capacities. In both cases, the relevant lower-level capacities, combined with pressures of the selective regime, instantiate a higher-level capacity of the population as a whole. It thus is plausible that there are no cases of selection to which the theory of systemic functions cannot apply—indeed, no cases of systemic change that cannot be explicated in terms of component capacities. However, the converse relation between explanatory strategies does not hold. There are cases of systemic instantiation that cannot be explained by appeal to mechanisms of selection. I have mentioned one already: A population may instantiate the capacity to evolve in the absence of selection so long as some other evolutionary force occurs. Random drift is one such force. But there are additional cases in which we explain the instantiation of some higher-level capacity by appeal to mechanisms other than those that drive selection. The capacity of the circulatory system to deliver nutrients throughout the body is instantiated, in part, by the heart's capacity to pump blood; this is a fact about the way the circulatory system works, not a fact about its history or its origins. Assembly lines instantiate the line's capacity to produce even when selection is absent. And consider the capacity of salt to dissolve in water. We explain this capacity by appeal to the bonding capacities of certain kinds of molecules. This is to appeal to mechanisms of constitution, not mechanisms of selection. The theory of systemic functions, therefore, having wider scope, renders the theory of selected functions redundant. And to the extent functions arise in systems affected by selection, these are best understood from within the theory of systemic functions.

III Virtues of This View

Besides the redundancy of selected functions, there are positive reasons for adopting the theory of systemic functions and dispensing with the theory of selected functions. I now describe some of the most obvious virtues of this approach.

Systemic Functions Are Basic

We should dispense with the theory of selected functions, construed as an autonomous theory, because systemic functions are more basic in two ways. The first is ontological. Selected functions are, at best, nothing more than systemic functions that also contribute to their own selective success. Systemic functions exist antecedently and, depending upon the demands of the selective regime, are selectively advantageous, selectively disadvantageous, or selectively neutral.[6] Traits preserved in the selective process by virtue of the selective efficacy of their systemic functions may, in consequence, acquire additional systemic functions relative to the selective success of the larger system. This extends the trait's causal efficacy outward to include contributions that increase the organism's reproductive output and thereby contribute to the population-wide redistribution of genotypes. Thus, for example, the heart has the systemic function of pumping blood, but also the function of contributing to the circulation of nutrients (by pumping blood), the function of contributing to greater reproductive output (by contributing to the circulation of nutrients by pumping blood), and, under appropriate conditions, the function of contributing to the redistribution of certain genotypes in the population (by contributing to differential reproduction by contributing to the circulation of nutrients by pumping blood). Likewise for other traits. We extend the layers of systemic functions by embedding the relevant trait and its most proximate function in the next higher level of organization, including the population and its capacity to evolve.

6. Preston (1998) presses this basic point but defends the thesis that the theory of selected functions ought to be retained along with the theory of systemic functions. By contrast, I take this point as compelling reason to discard selected functions in favor of systemic functions, including systemic functions that arise in systems affected by selection.

Some advocates of selected functions—notably, Neander (1995)—claim that these sorts of stacked functions qualify as autonomous selected functions. I disagree. Such functions clearly are systemic functions first, being the material upon which selection acts. If selected for, they may subsequently meet some of the conditions laid out in the theory of selected functions. But this is to say merely that they then become the sorts of systemic functions that contributed to evolutionary change (or stasis) via selection; they are systemic functions the effects of which emanate outward to the capacities of several larger containing systems, including the population. Selected functions are selectively efficacious systemic functions; there is no difference in the kind of function involved, only a difference in the scope of the system under investigation.

In a moment of characteristic humility—in the chapter "Difficulties on Theory"—Darwin, in *On the Origin of Species,* confesses the apparent absurdity in imagining that organs as complex and adapted as the eye could have evolved via natural selection (see the second epigraph to this book). He is not cowed, however, by the appearance of absurdity. He begins his response by insisting that reason—as opposed to immediate impression—tells him that the slow, gradual effects of selection could give rise to complex and adapted traits. He thus counters the impression of absurdity with a cautious but empirically grounded how-possibly argument. Part of the argument appeals by analogy to traits of other organisms that appear to have altered their functions over time. For example:

> Two distinct organs sometimes perform simultaneously the same function in the same individual; to give one instance, there are fish with gills or branchiae that breathe the air dissolved in the water, at the same time that they breathe free air in their swimbladders, this latter organ having a ductus pneumaticus for its supply, and being divided by highly vascular partitions. In these cases, one of the two organs might with ease be modified and perfected so as to perform all the work by itself, being aided during the process of modification by the other organ, and then this other organ might be modified for some other and quite distinct purpose, or be quite obliterated. (Darwin 1859, 190)

Organs endowed with a function can be preserved by selection when the functional effects are selectively efficacious. Organs endowed with a function can also be altered or eliminated by selection when the functional effects are overridden or when they are selective failures. Either way, selection acts upon existing traits and existing functional effects. And in the case of gills and swimbladders, it is clear that selection preserves one organ because of the selective efficacy of its systemic function and then alters or eliminates the other organ insofar as its systemic function becomes relatively inefficacious in its selective consequences. Selection, in short, acts upon the raw material of systemic functions, culling and disposing some, letting others stand.

The second way in which systemic functions are basic is epistemological. We cannot discover the selected function of any trait without first knowing its systemic function. If we do not know the systemic function of a trait, we have no guide with which to seek historical evidence for the claim that this trait was selected for the specified functional task.[7] This point is well illustrated by Kingsolver and Koehl (1985).[8] Their aim is to explain the evolution of insect wings. We know from Paleozoic fossils that the wings of early insects were much smaller and thicker than those of more recent descendents. Kingsolver and Koehl first hypothesized that these proto-wings were selected for lift, glide, even flight. To test this, they constructed physical models based on fossils and placed them in a wind tunnel, only to discover that the wings were too small to produce any lift. They concluded that the proto-wings could not have been selected for flight. They then hypothesized that there was selection for thermoregulatory effects, since the tissue of fossil wings appears to be a good conductor of heat. Their tests of physical models revealed that, indeed, the thicker proto-wings would have been efficient heat conductors, which would have kept the organism warm for longer periods of time, thus increasing the quantity of activities, including reproducing. They concluded that proto-wings could have been selected for thermoregulation. And, they conjectured, with eventual mutations for larger size, later versions could have been selected for flight as well.

This is a plausible theory of the evolution of insect wings. It is plausible because the conclusions drawn—conclusions concerning the possible causes of evolutionary change—are supported by empirical tests. This gives the conclusions some degree of probability. But, of course, the

7. I press this point in Davies (1996) and (1999) as it applies to the methods of evolutionary psychology.

8. Brandon (1990) discusses this example in the context of demonstrating the importance of how-possibly explanations in evolutionary theory.

relevant tests are compelling only because Kingsolver and Koehl knew so much about the organism and its powers in advance. They knew the size and thickness of the proto-wings, the relative size of wings and body, the tissue of the wings, and so on. On the reasonable assumption that the laws of physics (aerodynamics and thermodynamics, in particular) have not altered since the Paleozoic, the results of their tests are hard to resist. This case illustrates the epistemic priority of systemic functions. Had Kingsolver and Koehl not had information provided by fossils, they would have lacked important parameters with which to construct their experiments. Their conclusions are convincing only because they could reconstruct conservative and testable hypotheses concerning the systemic functions of the proto-wings. Information concerning systemic functions enabled Kingsolver and Koehl to offer a compelling how-possibly explanation of the evolution of insect wings.

Of course, dispensing with the theory of selected functions does not absolve us from the considerable epistemological obstacles that hinder the attribution of systemic functions to systems affected by selection in the past. The difficulty is not simply the paucity of relevant historical records. The more specific difficulty, which attends the study of selection even in contemporary populations, is that we often cannot determine which traits are being affected directly by selection. The attribution of a systemic function produced by selection involves the attribution of a specific task to a specific trait. This requires evidence that tokens of this particular trait—not some other trait with which this one is connected in some way—are responsible for the selective response. It also requires that tokens of this trait are responsible by virtue of performing this specific task, not some other. Yet often times genetic and developmental mechanisms affect various phenotypic traits within an organism so that several traits respond all at once to a given selective pressure.[9] Adducing evidence that supports the attribution of any such function is no easy task. I return to this point in chapter 5.

So, those features of selected functions worth preserving are best seen as systemic functions that contributed to selective success. Moreover,

9. See Amundson and Lauder (1994) for this point and related difficulties. See also Davies (1996).

knowledge of systemic functions is more basic, since we cannot trace the selective history of a trait unless we know its systemic effects among ancestral organisms. And, as argued above, the theory of selected functions construed as an autonomous theory is redundant, since systemic functions that emerge in systems affected by selection capture everything that the theory of selected functions can capture.[10] There is, therefore, no compelling reason to retain both theories.

This view has the further advantage of putting enthusiasm for adaptationism into perspective. Adaptationism in evolutionary biology is ambiguous between a substantive and a methodological thesis. The substantive thesis asserts that most or all organismic traits are the products of evolution by natural selection, while the methodological thesis asserts that, in trying to understand a given trait, we should assume it to be a product of natural selection, since that provides our best guide to discovering the trait's function. Advocates of selected functions are cheered by the pervasive influence of either version of adaptationism, since one appears to justify the attribution of selected functions on methodological grounds and the other on substantive grounds. But such optimism is unwarranted. If the thesis of this chapter is correct, substantive adaptationism is the claim that all or most traits have systemic functions that have been selected for, and methodological adaptationism is the claim that we should assume that all or most traits have systemic functions that have been selected for. I have doubts about both versions. The point, however, is that construing selected functions as systemic functions does not diminish the evidential demands that burden the adaptationist program. It shows, rather, that implementing the adaptationist's program gives neither aid nor comfort to the theory of selected functions, since the theory of systemic functions fits the bill just as well.

The same point holds for what Griffiths (1996) calls the "historical turn" in the study of adaptation—the confirming or disconfirming of adaptationist hypotheses by way of phylogenetic tests. Advocates of selected

10. Not everything that advocates of selected functions *wish* to capture; only what their theory in fact *can* capture. Advocates wish to account for the possibility of malfunctions. But, as noted above, the theory of selected functions lacks the resources with which to do so. See chapter 7 for the full argument.

functions may greet Griffiths's suggestions with enthusiasm, assuming that the sorts of tests he describes dissolve the evidential worries about attributing selected functions. But that assumption is false. Rather, the tests described provide some (though sometimes very little) information about the sorts of systemic functions that were selectively successful. The theory of systemic functions thus meshes well with the claims of evolutionary biologists: the theory of selected functions can claim no especial kinship to evolutionary theory.

Theoretical Unification

My view that selected functions ought to be discarded in favor of systemic functions provides theoretical unity that Kitcher's attempt at unification fails to provide. Kitcher suggests marrying the theory of selected functions with the theory of systemic functions under a general rendering of the concept "design": "This unity is founded on the notion that the function of an entity S is *what S is designed to do*" (Kitcher 1993, 379). The design in artifacts originates in the intentions of the designers, producers, or consumers; the design in natural objects originates in the processes involved in evolution by natural selection. This attempt at unity thus commits Kitcher to the claim that evolution by natural selection produces "design" of some sort or other.

But I take it that genuine unification in this case requires a single, nondisjunctive concept of "design" that applies equally to artifacts and to the products of evolution due to natural selection. And there is no such concept. It is of course plausible to describe the processes that lead to the production of artifacts as involving design; if not, it is hard to see what the concept "design" comes to. But it is not plausible to say the same about the processes that lead to evolution via selection of natural objects, no matter how many marks of apparent design they bear. The theory of evolution by natural selection helps explain why the biological realm appears as if it were designed. It does not explain, nor does it entail an explanation of, why the biological realm was designed, for no such process ever occurred (except, of course, in cases of genetic engineering, etc.). In consequence, there is no one concept of design to which Kitcher can appeal. Rather, there are two concepts quite distinct from one another, one applying to processes involving genuine design ("design"), the other to processes involving no design at all but merely processes that produce sufficient regularity and systemic complexity to give the appearance of design ("bearing-apparent-marks-of-design"). And if there is no single concept "design" that applies to both artifacts and natural objects, the proposed unification fails.[11]

Kitcher asserts that "one of Darwin's important discoveries is that we can think of design without a designer" (1993, 380). But this is ambiguous. It may assert merely that (a) Darwin showed us how to explain the apparent marks of design without postulating a designer. Alternatively, it may assert that (b) Darwin showed us that there really is design in the natural world despite the absence of a designer. Now I take it that (a) is true. But as we have seen, the truth of (a) cannot support Kitcher's proposed unification, since the concept "design" applies to the production of artifacts while the quite different concept "bears-apparent-marks-of-design" applies to the evolution via selection of natural objects. I also take it that (b) is false. For no one, not even Darwin, can show that there is design in the absence of a designer, since no one can refute the rather plausible conceptual claim that design requires a designer. I will say more about this conceptual claim momentarily. In the meantime, and conceptual claims aside, it is just not true that Darwin thought there could be design with no designers. On the contrary, Darwin's view is that we can explain away the apparent marks of design without appeal to a designer.[12] And to explain away the appearance of design without appeal to a designer is not to show that we can think of design without a designer; it is rather to show that we ought to cease thinking of the natural world as designed. So Kitcher's proposed unification cannot succeed.[13]

11. No doubt we can mock-up disjunctive concepts of design, but genuine unification is achieved only when the unifying concept is itself sufficiently unified. I take it that Kitcher agrees with this assumption. See his discussion of spurious unification in section 8 of his (1981).

12. See Hull (1973), especially pp. 55–66. See also Gillespie (1979).

13. Allen and Bekoff (1995) purport to "naturalize" our concept of design along with our concept of function. Although the considerations offered are interesting, the claim that they are "naturalizing" our concept of design is doubtful. So far as I can see, they are eliminating the concept of design from the biological realm, for they explicate design in terms of a quite different concept. This further concept has some analogies to our concept of design, but the disanalogies are obvious

What makes Kitcher suspect that there is a single concept of design applicable to artifacts and natural objects alike? He is moved by the example, described in chapter 2, in which a screw falls accidentally into the workings of a machine while it is being assembled. Despite its unintended role within the machine, this accidental screw nevertheless has a function derived from the intended purpose of the machine as a whole—or so Kitcher claims. Since there can be unintended features of artifacts that are functional nonetheless, there must be a concept of design applicable to artifacts that does not appeal to the direct effects of a designer. The same point applies to natural objects: They can have functions even if they have not been directly selected for, on the grounds that they serve some function derived from the naturally selected purpose of the larger system within which they work.

I agree that not all features of artifacts are the direct products of designers' intentions. I also agree that in some cases it may be tempting to attribute a function to such unintended features. But this hardly shows that there is a single concept "design" that applies to artifacts and natural objects alike. At least two considerations are relevant here. First, we should not agree that the accidental screw has any such function, no matter how tempting. We can trade intuitions and temptations all we like, but the bottom line is that this sort of example generates intuitions far too weak to do any substantive theoretical work. So we should reject the conclusion Kitcher wishes to draw from this example.

Second, even if we grant that the accidental screw has a function of some sort, it does not show that we are applying a concept of design that does not require a designer. Perhaps we are employing a notion of functions that makes no appeal to considerations of design; perhaps, for example, we attribute a function to the screw on the grounds that it satisfies the conditions put forth in the theory of systemic functions. Or perhaps we attribute a function on the grounds that, had the screw not fallen into that very place, the designer eventually would have placed a screw (or some functional equivalent) there anyway. I am not asserting that this is

---

and significant. I am in favor of eliminating our concept of design from the realm of nonengineered natural traits, where it has no proper application. But eliminating the notion leaves nothing for us to naturalize.

true, only that it is a possibility that Kitcher's example does not rule out. And notice this is not true in the attribution of functions to natural, non-engineered objects. We do not attribute a function to any natural object on the grounds that, if it had not evolved, then some designer or some surrogate process would have placed a functionally equivalent item there. For that may be false. If the relevant item had not evolved, then perhaps some other trait would have filled in, but it is also possible that the larger system would have been selected against or that the species would have gone extinct.

Moreover, the fact that most features of artifacts are the direct products of our intentions is obviously relevant, since none of the features of natural objects is a product of intentions. The case of the accidental screw does not alter this basic fact. The accidental screw shows that, for any *artifactual* system, a particular component of the system may acquire an unintended function. It does not follow that intentions are not necessary for artifactual functions. Indeed, intentions fix the functions of the whole system and the accidental screw derives its function from these. Were it the case that nothing fixed the functions of the whole, or at least the functions of major subsystems, the accidental screw would be functionless. And the obvious point is that, while intentions fix the functions of artifacts, nothing fixes the functions of *natural* systems. Natural systems bear apparent marks of design, so perhaps some natural traits, given their contributions to the larger system, derive their apparent marks of design from the apparent design of the system. But the disjunction of apparent and genuine design is a contrived concept lacking the requisite sort of unity.

Perhaps Kitcher could regroup. As formulated, the unifying concept seems to refer to the *process* of design. The function of an entity, he claims, is to do what it "*is designed to do*" (1993, 397). But suppose we focus not on the process of design, but rather on *having* a design, where the latter does not require the former. The suggestion is that there exists an act/object ambiguity in our concept of design and that my criticism fails to consider design as an object.[14] It may seem plausible, for example, that the mammalian eye has a quite definite design even if we grant that

14. I am grateful to Robert Cummins for suggesting this alternative.

evolution by natural selection is not a designing process. It has a design by virtue of its constitution—by virtue of internal complexity, adaptedness, efficiency, or some such property—not by virtue of its mode of production. Call this "constitution design." By contrast, an object may be designed even though it lacks internal complexity, efficiency, etc., so long as it was produced in the right way—in accordance with some intention. Call this "production design." We thus could reformulate Kitcher's unification this way: The function of S is whatever S was designed to do (production design) *or* whatever S is supposed to do given the design of the larger system (constitution design).

This reformulation helps only if the proposed notion of design is sufficiently unified. But it is not. The attempt to combine production and constitution design into a single concept conflates two importantly distinct notions that ought to be kept separate. The notion of production design is normative. As Kitcher claims, artifacts have functions by virtue of the intentions of designers, producers, or consumers. And components of artifacts like the accidental screw have unintended functions only if the larger system has at least one intended function. So, the functions of artifactual systems and their components, being the products of our intentions and being the objects of various social, moral, and legal conventions, plausibly possess norms of performance. But the second notion—constitution design—is not normative. At the very least, advocates of selected functions and advocates of the combination approach owe us an *account*—rather than the mere assertion—of how the processes in natural selection give rise to norms of performance. I doubt that any such account is forthcoming.[15] So, the two types of design differ substantially. The case of the accidental screw does not diminish this difference.

As an aside, we may wonder whether we ought to preserve the notion of constitution design. We may wonder about the theoretical utility of conceptualizing the mammalian eye in terms of constitution design even while acknowledging the difference between constitution and production design. This may seem a minor issue. So long as we do not fall back into conflating the two notions, describing the eye in terms of design appears unobjectionable. But, minor or not, two reasons tell against applying *any*

15. I defend these doubts in section III of chapter 5 and in chapter 7.

notion of design to natural traits. The first is that the history of the debate over functions inclines many theorists to conflate the two notions. Wright (1973) explicitly aims to provide an account that applies to artifacts and natural traits alike. So does Millikan (1984). And two decades after Wright, Kitcher (1993), Griffiths (1993), and others are pursuing a similar goal. Since the terms of the debate incline many of us in the wrong direction, perhaps a bit of linguistic legislation is in order. Second, there are no good theoretical grounds for describing natural traits in terms of design, including constitution design. This is because the theory of systemic functions provides resources with which to describe both the evolution and the detailed workings of the eye without any mention of design. The attribution of systemic functions at the population level explains how the category of eyes evolved as it did; attributions at the level of token eyes explain how they do what they do. That, I submit, is explanation enough. Talk of design adds nothing in the way of information or conceptual clarity, and only inclines us to persist in seeing the natural realm falsely in terms derived from the artifactual.

Kitcher's attempted unification, therefore, cannot succeed. The unity afforded by my view, by contrast, does not rest upon the dubious claim that there is a single concept of design applicable to artifacts and natural objects alike. Surely this is an advantage. If unity is in fact a virtue in scientific explanation—as Kitcher suggests and as I agree—then my view is preferable. My view is also preferable to either the competitor or the complement view on grounds of unification, since neither offers any hope of unification. Indeed, if theoretical unification is a worthy pursuit, then the view offered here wins hands down, as it covers function attributions in the biological and the nonbiological sciences, and does so without asserting that Darwin somehow saved the concept of design while gutting it of all content.

The Duration of Functions

A further virtue of my approach, as against the historical approach, is that the theory of selected functions must grapple with issues of temporal relevance that raise no problem for my view. The theory of selected functions must answer the question, If functions are a product of selective history, then precisely which stretch of history is relevant to the determination of a trait's present functions? After all, at one time the human appendix probably was selected for some of its causal effects, which means we can explain why the appendix has persisted by appeal, at least in part, to its past selective history. In consequence, at least some versions of the theory of selected functions—for example, Millikan (1984)—are saddled with the unhappy burden of attributing selected functions to vestigial traits such as the appendix. In response, and in an attempt to improve the theory, Griffiths (1992) and Godfrey-Smith (1994) claim that only the most recent period of past selection is relevant to the determination of current functions, thereby excluding the distant selective history in which the appendix proliferated. Alternatively, Kitcher (1993) claims that the most recent period of past selection and current selection are both relevant, thereby excluding distant periods in which the appendix proliferated and including current periods in which the appendix is not selected for. On my view, however, this entire question is a nonissue. The attribution of systemic functions to a population in which selection occurs is always relative to a specific set of selective demands and thus to a specific stretch of selective history. Time-relativity is integral in attributing systemic functions to systemic capacities; the question concerning which stretches of selective history are relevant is answered in the course of determining an item's selected systemic function. We thus can claim of the appendix that it used to have a systemic function that it has since lost—and that seems just right.

Preserving Distinctions

My view also addresses the distinction, marked by Mayr and Tinbergen, between evolutionary explanations (in terms of historical causes) and functional explanations (in terms of proximate, systemic causes). As we have seen, some philosophers suggest that the prevalence of this distinction within biological practice provides reasons for thinking that each theory of functions applies to distinct theoretical phenomena and hence that these theories are complements rather than competitors. I disagree. The prevalence of this distinction provides reason for thinking that there must be some difference or other in the sorts of explanations we employ, but it does not provide reason for thinking that this difference must be between explanatory strategies or forms of explanation. On my view, the difference to which Mayr and Tinbergen refer is nothing more than a difference in the kind of system we are trying to explain. The object of explanation differs, but the form of explanation is the same. From the perspective of the theory of systemic functions, Mayr's and Tinbergen's evolutionary explanations involve systemic functional analyses of populations and their capacity to change by virtue of their own powers. The functional explanations described by Mayr and Tinbergen likewise involve systemic functional analyses, but analyses of more or less discrete systems (typically organisms) and their various capacities, rather than entire populations. We can, therefore, preserve these important theoretical distinctions from within the theory of systemic functions. The theory of selected functions adds nothing.

We also can agree with Godfrey-Smith that adequate explanations should uncover rather than conceal causal information and nevertheless insist that the theory of selected functions is extraneous to this task. The theory of systemic functions, applied to populations, reveals all the causal information uncovered by the theory of selected functions. In fact, the theory of systemic functions reveals more information, since it, unlike the theory of selected functions, is free to assign functions that arise as a consequence of nonselective forces. But even if we restrict our attention to the effects of selection, the theory of systemic functions, applied to a population's capacity to evolve, requires us to distinguish between causal effects that contributed to selective success and those that did not. There is no loss of relevant information.

IV Related Views

The recent literature on functions contains at least three views that, judged from a distance, appear similar to mine. Griffiths (1993), Walsh and Ariew (1996), and Buller (1998) represent a shift away from historical functions toward the systemic capacity approach. All three assert that selected functions should be construed as a specific kind of systemic function—systemic functions produced by selection—and in this regard these views are somewhat near my own. Nevertheless, these views aspire to retain key features of the theory of selected functions and that, in my view, is a mistake. As I now shall argue, my account of systemic functions represents a more radical break with the historical approach, and that makes my view preferable.

The first point of difference is step (1) of the strategy described in section II above. The difference here may be nothing more than a matter of completeness or explicitness, but it seems to me important all the same. Of the three alternatives, none requires a systemic analysis of the population; all three focus immediately upon the organism. Griffiths, for example, focuses directly upon rates of reproduction caused by selection. He asks, What are the salient capacities of the various parts of this organism such that tokens reproduced at this rate? On this alternative, we begin with step (2) of my strategy, which takes us immediately to traits that answered the demands of the selective regime. We then proceed to step (3) if we are interested in traits operating at lower levels of organization. Walsh and Ariew advocate a similar view, as does Buller, though Buller focuses upon contributions to organismic fitness rather than selection. But on any of these alternatives, step (1) of my strategy is omitted or at least unacknowledged, in which case my view is to be preferred. For, evolutionary biologists are interested in the properties of populations—in the ways they change or resist change—and my view places the workings of organismic traits squarely within the workings of the population as a whole. This deepens our understanding of the origin and nature of functions that arise in systems affected by natural selection.

A more substantive difference concerns the status of malfunctions. The three alternatives agree that selected functions are a kind of systemic function. But they hold that selected functions, despite being instances of systemic functions, are nevertheless normative in character and thus account for the possibility of malfunctions. All three are quite explicit on this point. Griffiths, for example, asserts that there are two notions of function within contemporary biology, one of which refers to the alleged purposive or teleological properties of traits. In this respect, these three views follow Bock and von Wahlert (1965) in holding that we need an account of the purposes or norms employed in biological theories. But I find this puzzling. It is widely agreed that the theory of systemic functions does not warrant the attribution of malfunctions.[16] None of these theorists, at any rate, argues to the contrary. Why, then, do they allow that selected functions, cast from within the theory of systemic functions, warrant the attribution of malfunctions? If systemic functions cannot account for malfunctions, and if selected functions are nothing but systemic functions, then selected functions cannot account for malfunctions either. Or, at minimum, a compelling argument is required to show that selected functions, understood naturalistically, are somehow unique among all systemic functions.[17]

There is, in consequence, a tension internal to each of these alternatives. Of course, advocates of these alternatives can deny any such tension and claim that systemic functions that contribute to selection are normative precisely because they contribute to selection. Selection, they might insist, gives rise to norms of performance that somehow transform mere systemic functions into functions endowed with normative powers. This seems to be just beneath the surface in Griffiths's discussion. But it is mysterious how selection can give rise to any such norms. Natural selection can cause the persistence or proliferation of trait types, but this involves nothing more than various causal-mechanical processes operating over two or more generations. It is implausible that the processes that give rise to systemic functions in systems affected by selection somehow produce purposes or norms of performance, while the processes that give rise to functions in systems not affected by selection do not. I agree that appeals to selection help us explain how the functional trait persisted, but that is just to say that there are causal-mechanical processes that resulted in the trait being preserved in the population. It is hard to see,

16. See, for example, Amundson and Lauder (1994), the main title of which is "Function Without Purpose." But see Godfrey-Smith (1993) for dissent on this point. Chapter 7 offers a rebuttal to Godfrey-Smith's suggestion.

17. As a matter of fact, I believe that neither theory of functions warrants the attribution of malfunctions. I defend this claim in chapters 6 and 7. I also believe that the theory of systemic functions offers a plausible speculative explanation for our evident inclination to (mistakenly) attribute malfunctions. I develop this idea in chapter 6. So, I do not accept the premises of the argument I am now offering. In particular, I do not accept that selected functions are special or unique among systemic functions. Advocates of these alternatives, however, do accept these premises, and that is a source of the internal tension noted here.

from a naturalistic perspective, how or why we should think that the processes that result in trait preservation confer upon descendent tokens a norm of performance. I return to this point in chapter 5.

It is easy to be seduced by our slightly metaphorical ways of describing matters. It is easy to slide from "T was selected because it did F" to "T was selected for F" to "T is for the sake of F." As naturalists, we must measure the value of our metaphors by the findings of our best-developed scientific theories. Thus, we must consider the available naturalistic grounds for claiming that selection transforms mere systemic functions into normative functions. Pointing out that some systemic functions—those that contributed to the evolution or stasis of the population—help us explain a trait's persistence in the population hardly seems sufficient. At the very least, the burden rests with advocates of selected functions to specify exactly how the processes involved in natural selection give rise to norms of performance. I doubt that can be done.

This internal tension aside, the more important point is the extent to which my view differs with respect to malfunctions. Systemic functions, on my view, are not simply more basic than selected functions. They are also the only kind of functional property. There are, of course, various instances of this general kind, including those that arise in systems affected by selection, but the theory of systemic functions, being the genus, dictates the features that any species can possess. In particular, no species can possess properties that conflict with the definition of the genus. Now the genus—the theory of systemic functions—asserts that item I has function F only when it is capable of performing task F. Loss of the capacity to perform F entails loss of functional standing. In consequence, no instance of the theory of systemic functions can possess any property that violates this condition. In particular, no instance can malfunction.

Of course, a genus is characterized in terms of properties that are general, while a species will likely possess specific properties not included in the description of the genus. Welsh Corgis comprise a category within the Canine genus and yet Corgis have a tendency toward obesity not shared by dogs in general. So the genus, in dictating to the species, does not dictate *everything*. But it *does* dictate with respect to those properties integral to the definition of the genus. No species can have a property that conflicts with properties essential to, definitive of, or in some way integral to the genus. And since systemic functions *require* possession of the capacity to perform task F, no token can have a systemic function in the face of physical incapacitation. This suffices for the claim that the instantiation view should discard rather than struggle to preserve the alleged normative feature of selected functions.

This, then, is why I reject the views of Griffiths and others. They attempt to preserve the alleged normative properties of selected functions from within the theory of systemic functions. That is a mistake. That is to assert that there can exist systemic functions that lack requisite physical capacities even though possession of systemic functions requires possession of requisite physical capacities. On my view, by contrast, since systemic functions cannot account for malfunctions, systemic functions that emerge in systems that contribute to selection likewise cannot account for malfunctions. Being the target of selection does not transform the basic character of a trait's function but merely extends the efficacy of the trait's effects to include the population and its capacity to evolve or remain static. We should reject those features of selected functions that conflict with the conditions integral to the existence of systemic functions.

The third difference also is substantive. The alternative views agree with one another that selected functions are systemic functions, and I hold that some systemic functions arise in systems affected by selection. This, however, puts considerable burden on the theory of systemic functions—it had better be a persuasive and promising theory on independent grounds. But, of course, critics of systemic functions claim that the theory is neither persuasive nor promising. One of the most prominent objections is that the theory is promiscuous in the functions it attributes. Millikan (1989), for example, complains that Cummins's (1975) formulation of the theory is unacceptable because it permits the attribution of systemic functions to rain clouds. Clouds, she asserts, apparently have the systemic function of putting moisture in the soil and thus contributing to vegetable growth, and this conflicts with the powerful intuition that clouds have no natural purpose. Millikan allows that clouds and other items may *function as* this or that, but she denies that they *have* any genuine functions. So, the theory of systemic functions runs together genuine and nongenuine functions; hence its promiscuity. If systemic functions are promiscuous in this way, and if selected functions are not promiscuous (as advocates of selected functions claim), then, prima facie, we should retain the theory of selected functions as a separate autonomous theory.

Thus, anyone claiming that selected functions are nothing more than one sort of systemic function must fend off the worry about promiscuity. Griffiths does not address this worry. Nor do Walsh and Ariew. And, although Buller faces the objection, his response is overly narrow. He restricts the attribution of weak etiological functions to traits in systems that have been subject to selection, and this is supposed to block the indiscriminate attribution of such functions.[18] This is a strategy available to all three alternatives: They may insist that systemic functions are not promiscuous precisely when they arise as a result of selection. But this restriction does not go far enough. The issue is whether selected functions should be assimilated to the theory of systemic functions. Unless the latter theory can be shown to be acceptable generally in *all* of the domains to which it purportedly applies—*including* domains in which selection does *not* occur (the workings of assembly lines, the constitution of salt, etc.)—we ought to resist any such assimilation. We need an argument for the claim that systemic functions discriminate in the right way across all applicable domains.[19]

By contrast, and as I shall argue shortly, the worry about promiscuity has no force against my version of the theory of systemic functions. I shall defend the following three theses. First, the promiscuity objection rests upon a pervasive but mistaken assumption. The assumption is that some effects of natural objects are genuinely functional—in the sense that such functions entail norms of performance intrinsic to the relevant traits that underwrite the possibility of malfunctions—while other effects are not genuinely functional but, at best, merely useful. I reject this assumption because I reject the claim that natural objects possess intrinsic norms

18. Buller's appeal to selection differs importantly from those made by most (if not all) advocates of selected functions. See especially section 3 of his discussion. The details of his view, while nuanced and compelling, are not to the point of this discussion.

19. To be fair, Buller is concerned to defend a version of the theory of etiological functions, not the theory of systemic functions generally, as the title of his paper makes clear. Nevertheless, the point remains that unless the theory of systemic functions is defended against charges of promiscuity, the assimilation of selected to systemic functions is far less attractive.

of performance. To accept the postulation of such norms is contrary to a naturalistic orientation in inquiry and certainly contrary to the methods and postulations of the theory of evolution by natural selection. I defend this claim in chapter 5. Second, while I deny that any natural objects are intrinsically functional in this sense, I concede that some objects—those with sufficient internal complexity—nevertheless seem to us more functional than others. And this appearance can be explained in terms of the theory of systemic functions. Part of that explanation requires a third thesis, namely, that the theory of systemic functions applies properly only to systems that are hierarchically organized. I defend these latter theses in chapter 4.

V Conclusion

I conclude, then, that we should embrace the theory of systemic functions and discard the theory of selected functions construed as an autonomous theory. Functions are effects that contribute to more-general systemic capacities. Such functions can arise in systems affected by selection, but only insofar as they contribute to the systemic capacity to change or resist change. Functional properties are constituted by contributions to systemic capacities, not by selection. I further conclude that my formulation of this thesis is preferable to other theses extant in the literature that may appear similar in structure to mine. Also, while I acknowledge that the theory of systemic functions has its critics and its flaws, I believe those critics can be fended off by developing the theory and redressing its flaws. I turn now to defend the theory against the charge of promiscuity.

4

Defeating Promiscuity: Hierarchical Systems and Systemic Functions

Of all the criticisms leveled against the theory of systemic functions, perhaps the most damning is that all too often the theory misses the mark. The theory licenses the attribution of functions when in fact no such functions exist. Or so it is claimed. Millikan (1989), for example, alleges that the theory warrants the attribution of functions to rain clouds, thereby jarring our intuitions concerning the norms of nature. Matthen (1988) claims that the unwieldy tusk of the narwhal whale has the systemic function of reducing the animal's mobility, jarring our intuitions further. Hence the charge of promiscuity. If the charge were true, the thesis of the previous chapter would be objectionable. The thesis of chapter 3 is that there is only one true theory of functions—the theory of systemic functions. On this view, selected functions are best discarded in favor of systemic functions. But advocates of selected functions claim that their theory, unlike the theory of systemic functions, is not promiscuous. Other things being equal, we should not relinquish a nonpromiscuous account of functions in favor of a promiscuous one.

In fact, however, the charge of promiscuity can be defeated by developing the theory of systemic functions in a relatively simple way. It is usually taken for granted that a systemic function analysis can be given of any phenomenon so long as someone takes an appropriate interest in some of its capacities and so long as it can be construed as a system of some sort. Critics seize on this and mock-up "systems" of all sorts and then pretend to embarrass the theory by showing how it warrants the attribution of functions when in fact no such functions exist. The central thesis of this chapter, however, is that there is a rather obvious restriction on the sorts of phenomena to which the theory of systemic functions is properly applied that defeats the charge of promiscuity. The restriction is that the relevant phenomena must be hierarchically organized—in the sense described below in section II. With this restriction in place, Millikan's rain clouds and Matthen's tusks lose their critical force. I leave open the possibility that the theory of systemic functions requires additional conditions to handle all possible objections. My thesis here is the modest claim that restricting the theory to systems that are hierarchical immunizes the theory against what is thought to be a deadly ill. I also argue that Cummins's original formulation of the theory of systemic functions contains resources with which to diminish significantly the force of the promiscuity worry. In the end, I advocate a formulation of the theory that differs from Cummins's, but the power of his formulation has not, in my view, been properly appreciated.

The claims of this chapter, however, do not dissolve the promiscuity objection entirely. A rather vague unease still nags. Even granting my claim that restricting the theory to systems that are hierarchically organized blocks the alleged counterexamples, there remains the sense that the theory of systemic functions fails to explicate the real norms of nature. The sense remains that effects of some traits are genuinely purposive—"properly" functional and "for the sake" of some end—while others are merely useful. The residual worry about systemic functions is that they cannot account for this intuitive difference. The theory of systemic functions asserts that functions are systemic capacities that contribute to higher-level systemic capacities, but this is considered too crude to distinguish capacities that are genuinely purposive from those that are not. I address this lingering unease in chapters 5, 6, and 7. I deny that any natural trait is "properly" functional in the way described by critics of systemic functions. The claim that natural selection is capable of producing norms of performance—norms that confer genuine purposiveness upon natural, nonengineered traits—is, I argue, a claim we ought to reject. It is here that my view is revisionist in character: The notion of "proper" functions among natural traits cannot be sustained within a naturalistic orientation toward inquiry. If my defense of this revision in our concept is plausible, then the promiscuity objection is entirely without force.

This leaves us with one last issue. If we give up the currently pervasive assumption that some natural traits are "properly" functional while others are merely useful, we nevertheless must wonder why we are so tempted to see the biological realm in these terms. It is natural to wonder why we have the vague sense that some natural objects are more functional than others. Now, perhaps the origin of this sense is entirely cultural-historical. Perhaps our worldview is so permeated by our theological heritage that we now find it "natural" (second nature) to see natural objects on the model of artifacts. But I want to suggest, in good Humean fashion, that certain psychological capacities and limitations incline us to see some objects as more functional than others. This is not to rule out the cultural-historical speculation. Indeed, the cultural-historical story may be part of a good Humean story. Nevertheless, I shall restrict my speculations to the psychological. It is not my aim to argue for these highly speculative speculations, but rather to suggest by way of example how to diminish the intuition we have that much of the biological realm is purposive. I turn to these speculations in chapters 5 and 6.

I Promiscuity and Cummins's Constraints

The charge is that the theory of systemic functions attributes functions indiscriminately to traits that intuitively are nonfunctional. One way to formulate the accusation is to point out that the theory warrants the attribution of systemic functions to traits that reduce an organism's fitness:

> The narwhal has a large tusk, which actually reduces its mobility. Presumably the tusk is there in order to increase the conspicuousness of a narwhal during mating situations. Obviously, then, the [selected] function of the tusk is to increase the sexual attractiveness of the animal; certainly its function cannot be to reduce mobility. Yet, a Cummins-style functional analysis can be given of how the tusk reduces the narwhal's mobility. (Matthen 1988, 15)

The claim is that the theory of systemic functions appears to justify the attribution of functions that are prima facie unreal. It is just unintuitive that functions are effects that in some way diminish an organism. And yet, so long as we are sufficiently unencumbered by the sorts of systems and systemic capacities to which we apply the theory, systemic functions that diminish mobility or other organismic powers appear to be justified by the theory.

Millikan offers what appears to be a similar objection, claiming that the theory of systemic functions warrants the attribution of functions to traits that have no natural purpose:

> according to Cummins' definition it is, arguably, the function of clouds to make rain with which to fill the streams and rivers, this in the context of the water-cycle system, the end result to be explained being, say, how moisture is maintained in the soil so that vegetation can grow. Now it is quite true that, in the context of the water cycle, clouds function to produce rain, function *as* rain producers; that *is* their function in that cycle. But in *another* sense of "function," the clouds have no function at all, because they have no purpose. (Millikan 1989, 294)

Millikan puts her finger on what appears to be a key intuition driving the charge of promiscuity. *Having* a function, she supposes, is distinct from functioning *as* something. Trait T can function as an F-er—as a performer of F—simply because it happens to perform F. But T's functioning as an F-er neither entails nor is entailed by T's actually having the function of F-ing. Having the function of F-ing involves possession of an office or role that includes its own norms of evaluation; functioning as an F-er does not. In particular, if T has the function of doing F, then tokens of T are "supposed to" do F—doing F is what tokens of T are "for"—even when they are physically unable to do F. The standards of performance imposed by the selective efficacy of ancestral traits apply even in the face of physical incapacitation. Thus, "a diseased heart may not be capable of pumping, of functioning *as* a pump, although it is clearly its function, its biological purpose, *to* pump. . . ." (1989, 294) This is not true of an object that merely happens to do F.

But the upshot of Millikan's example is ambiguous. The point might be merely that we have two concepts of functions—one concerning the capacities of systemic components, the other concerning genuine natural purposes—and that we ought not run them together. That, however, is not convincing. I have already argued (in chapter 3) against the two-concept view of functions and, moreover, I give arguments against the postulation of "genuine" natural purposes in chapters 5, 6, and 7. But Millikan might intend something else. The point might be that the theory of systemic functions, insofar as it warrants the attribution of functions to rain clouds, is indiscriminate and thus unacceptable. The point might be that the theory of systemic functions is promiscuous, attributing functions to items devoid of natural purposes and hence devoid of real functions. Taken in this way, Millikan's charge is of a piece with Matthen's. The aim of this chapter is to show that this charge can be defeated.

Advocates of systemic functions, myself included, must face the claim that some natural items appear properly functional while others do not.

This I attempt in later chapters. Here, however, I focus directly on the claim that the theory of systemic functions warrants the attribution of functions to traits that in fact are functionless. This claim has an air of plausibility about it when we focus on the specific examples given, especially Millikan's rain clouds. In fact, however, these examples rest upon an assumption that we ought to reject. In the case of the clouds—and in other cases discussed below—it is assumed that the theory of systemic functions properly applies to any system and any systemic capacity we care to describe. It is assumed that the only constraints on the sorts of phenomena to which the theory applies are our explanatory interests or, better, our explanatory whims. But that is false. The theory, as I argue below, is properly applied only to systems that are hierarchically organized. With this restriction in place, it is no longer the case that the theory applies to just any phenomena in which someone may take an interest—and this makes the charge of promiscuity far less compelling.

Before defending this claim, however, I pause to review Cummins's original constraints on applications of the theory. Cummins anticipates the worry about promiscuity and offers compelling suggestions for resolving it. Consider a systemic analysis of the "circulatory noise" produced within all mammals:

> Each part of the mammalian circulatory system makes its own distinctive sound, and makes it continuously. These sounds combine to form the "circulatory noise" characteristic of all mammals. The mammalian circulatory system is capable of producing this sound at various volumes and various tempos. The heartbeat is responsible for the throbbing character of the sound, and it is the capacity of the heart to beat at various rates that explains the capacity of the circulatory system to produce a variously tempoed sound. (Cummins 1975, reprinted in Sober 1994, 65)[1]

This scenario is akin to Matthen's tusks and Millikan's clouds. We are disinclined to attribute to the heart the function of making a noise despite

1. The first sentence of this passage is false. Not every part of the circulatory system makes a noise. Or, more modestly, not every part makes a noise of any consequence. Perhaps it is true that capillaries emit noises in the process of transferring nutrients across cell membranes. And perhaps it is true that the brain, in producing the electrical signal that triggers dilation of the heart, also makes a noise of some sort. But neither of these would appear in a systemic functional analysis of the circulatory system offered, for example, by a practicing cardiologist. What this shows is that not every part of the circulatory system would appear in a systemic functional analysis of circulatory noise. I return to this point below.

the truth of each of the claims involved in the analysis. But we are disinclined, according to Cummins, not because heart sounds failed to contribute to ancestral fitness or to selection—what is the historical evidence that they failed?—but rather because the analysis of the mammalian circulatory system into such component effects is, from the point of view of scientific inquiry, relatively uninteresting and uninformative.

Cummins lays down three conditions for assessing the level of interest and informativeness of any such analysis. A systemic analysis is of scientific interest only if it

(i) analyzes the target capacity into capacities that are less sophisticated,
(ii) analyzes the target capacity into capacities that are different in kind, and
(iii) offers an analysis (a program or flowchart of components and their effects) the complexity of which is sufficient to bridge the gap between the target capacity and the analyzing capacities. (Sober 1994, 66)

Consider again the case of the assembly line. We analyze the capacity of the line to produce some product in terms of its component capacities and we analyze each component capacity into its component capacities and so on. If these latter capacities are relatively unsophisticated and different in kind from the line's capacity to produce its product, and if the analysis offered is sufficiently rich in causal information to bridge the gap, then the proposed analysis and concomitant attribution of functions are warranted. Otherwise not.

Critics of systemic functions have failed to appreciate the power of these constraints. Consider, first, the circulatory noise example. The system under investigation is (allegedly) the circulatory system and the systemic capacity we wish to explain is the production of circulatory noise. The heart, of course, is the main maker of noise, but that does not justify the attribution of any such systemic function to the heart. For, conditions (i) and (ii) require that the target capacity—the capacity of the circulatory system to make noise—be analyzed into other capacities that are simpler and different in kind from the target capacity. We cannot, therefore, attribute to the heart the function of making noise, since that capacity is neither simpler nor different in kind from the capacity of the circulatory system to make noise. The noise of the circulatory system, taken as a whole, may differ in subtle ways from the noise produced by the heart, but there is very little difference in degree and hardly a difference in kind. Both produce throbbing, thumping noises. And if the noise of the circulatory system has any complexity, then so too does the noise of the heart, consisting of the noises produced by expanding and the sound of blood entering the heart, and by contracting and the sound of blood exiting the heart. This example, therefore, fails to raise an objection to the theory of systemic functions because it founders on Cummins's constraints.

Now consider Millikan's clouds. What are the relevant system and systemic capacity? Presumably, the system is the water cycle. The systemic capacity, however, is not clear. Millikan describes it this way: ". . . the end result to be explained [is], say, how moisture is maintained in the soil so that vegetation can grow." The capacity, then, is either (a) maintaining soil moisture or (b) vegetable growth, or both. These are importantly different capacities, so I consider each in turn. Begin with (a). It is doubtful that maintaining soil moisture is part of the water cycle, since the cycle would persist in the absence of soil. But we can meet this problem by describing the capacity more generally. What is required for persistence of the water cycle is the existence of water on or near the surface of the earth so evaporation can occur; the existence of soil is incidental to this more general requirement. We thus can describe the capacity as (a*) the distribution of moisture on or near the surface of the earth.

If (a*) is the systemic capacity of the cycle, then it is plausible that Cummins's constraints dissolve Millikan's clouds. We must ask whether the analyzing capacity (the clouds producing rain) is simpler or different in kind from the target systemic capacity (the distribution of moisture). I am inclined to answer in the negative. The clouds producing rain just *is* the distribution of moisture on the earth's surface. At the very least, the clouds producing rain has the direct effect of distributing moisture, in which case the burden rests on critics of systemic functions to show how the analyzing capacity of the clouds is simpler or different in kind from the target capacity of distributing moisture. The burden is on Millikan to show that Cummins's constraints (i) and (ii) have been met. I doubt that can be done.

It might be retorted that, no matter what we think in this specific case, Cummins's (i) and (ii) are too abstract to give a clear answer in all cases.

This, I believe, is an objection that must be taken seriously. After all, it is not always obvious when one phenomenon is simpler than or different in kind from another. In general, advocates of systemic functions should not put too much stress on conditions (i) and (ii), for doing so may haunt them in cases where the theory appears to apply but where the assertion of greater simplicity or a difference in kind can be called into question. Such vagueness can cut for or against the theory of systemic functions. This is so, I believe, even though vagueness is not a problem in the case of the water cycle. I return to this point presently.

But now consider systemic capacity (b), vegetable growth. Cummins's constraints appear to have less force in this case, since the analyzing capacity (the clouds producing rain) is simpler than and different in kind from the target capacity (vegetable growth). But this is not enough to sustain Millikan's objection. The obvious problem is that the alleged target capacity—vegetable growth—is not a systemic capacity of the water cycle. A mere water cycle does not give rise to vegetable growth. This is a significant point. Two considerations are relevant. First, vegetation is not an integral part of the water cycle. The cycle includes evaporation of moisture and the production of clouds and precipitation, but not vegetation. The cycle would persist in the absence of vegetation and presumably predates vegetation.[2] Claiming that vegetable growth is a capacity of the water cycle is like claiming that the rattling of my kitchen window caused by the factory assembly line across the street is a capacity of the assembly line. Not every effect of a system is a systemic capacity (more on this in section II).

Second, if vegetable growth is a capacity of some system, the relevant system must be much broader in scope than the water cycle. This is a consequence of the theory of systemic functions. A necessary condition for the attribution of systemic functions is an analysis of the relevant system that "appropriately and adequately" accounts for the specified

2. This is not to deny that vegetation plays an important role in the water cycle as it exists today. Producing oxygen and absorbing carbon dioxide are capacities of great import to the existence of living organisms. But once we point out the importance of vegetation to organismic life, it is clear that the relevant system in need of analysis is much broader than the water cycle. See the next paragraph, as well as section III, for discussion.

systemic capacity (see chapter 2, section II). We begin by delineating the higher-level capacity we wish to explain and analyze downward until we reach the physical mechanisms instantiating the itemized capacities. And the capacities itemized must in fact account for the higher-level capacity being analyzed. It is obvious, however, that we cannot explain vegetable growth adequately by appeal to the water cycle, for the water cycle alone does not a garden make. We must also appeal to the genetic constitution of vegetables, the chemical composition of soil or water, the effects of ultraviolet radiation, the dispersal of seeds by wind and animals, and so on. Now, perhaps there is a system that includes all of these elements as components and perhaps relative to an analysis of that quite different system we can identify the systemic functions of the water cycle. And perhaps the function of the water cycle within that system is the production of rain and thus the distribution of moisture on or near the earth's surface. But if so, then once again Millikan's objection fails to satisfy Cummins's constraints. For if, relative to this much larger system that produces vegetable growth, the systemic function of the water cycle is to produce and distribute rain, then we cannot attribute to any part of the water cycle (clouds, in particular) the systemic function of producing and distributing rain. Doing so would flout conditions (i) and (ii).

I conclude, then, that the constraints offered in Cummins (1975) defeat the objection raised in Millikan (1989). Nevertheless, I do not think that Cummins's constraints are decisive. As indicated, Cummins casts his constraints in terms of explanatory information. Systems that are simple and unstructured exhibit little in the way of systemic functions and thus an analysis of such systems produces little in the way of explanatory information. We gain very little if told that a particular systemic capacity is the result of some component exercising that same or a similar capacity. Hence the plausibility of requiring that the analysis explicate the systemic capacity in terms of lower-level capacities that differ in kind and sophistication. Despite their plausibility, however, there are two problems with Cummins's (i) and (ii). The first, marked above, is that they are vague. Now, vagueness in itself is no objection if we can trace its source. But the requirement that the analyzing capacities be simpler and different in kind prompts the question, Simpler and different in kind *in what respects?* How are we to know when we have an analysis that satisfies (i) and (ii) in ways that are salient? We can answer this question in terms of the sorts of systems for which a systemic analysis is appropriate. And this brings me to the second problem with Cummins's constraints. The focus upon explanatory information, while undoubtedly important, is secondary. Of primary importance is the *kind of system* we are trying to analyze in terms of systemic functions. If the alleged system is simple and does not consist of distinct levels of organization, then the theory of systemic functions cannot be applied. This is because the theory requires the production of some higher-level capacity by the organized effects of lower-level capacities. The theory requires that the system under analysis be structured in the right way. Once we have an account of the relevant sorts of structure, Cummins's constraints concerning explanatory information fall out as a consequence. Such constraints, therefore, are not constitutive of the proper application of the theory, but a consequence of a core restriction on the sorts of systems to which the theory applies. I turn now to consider the relevant restriction.

II Hierarchy[3]

The sorts of phenomena to which the theory of systemic functions properly applies are those that are *hierarchically organized.* A system is hierarchical if it exercises a capacity at one level by virtue of the organized capacities operating at some lower level of organization. Lower-level capacities include the effects of *structural components* and *causal interactions* within the system. Among structural components there is little or no interaction; it is the organization among such structural features that contributes to the exercise of a higher-level capacity. The single fibers in Haugeland's fiber-optic cable, described in chapter 2, illustrate this sort of hierarchy. Such components possess structural systemic functions. However, lower-level capacities typically include, in addition to certain structural components, causal interactions between the effects of various

3. The discussion in this section is indebted to work on hierarchical systems in Simon (1969), Wimsatt (1986), and especially Bechtel and Richardson (1993). I have nothing original to add to their accounts. Nor is it my intention to do exegetical justice to their views. I merely intend to select key insights and import them to the general account of functions provided by Cummins.

components. Here the emphasis is on organized interactions rather than structural organization. The effects of the heart's pumping interact with various effects of the arteries and blood, the lungs and the brain, and so on. These interactions give rise to the circulatory system's capacity to deliver nutrients. Such components possess interactive systemic functions. A system, then, is hierarchical when some higher-level capacity arises out of the structural or interactive systemic functions among lower-level components.[4]

These two forms of hierarchy entail some degree of internal *systemic cohesiveness*. Systems that are structurally hierarchical are *structurally cohesive*, where a change in the organization of lower-level structural features affects the exercise of the higher-level capacity. Reorganizing individual fibers of Haugeland's fiber-optic cable will alter or destroy the transmitted image. Or, a change in the structure of a population will likely affect its capacity to evolve or its capacity to remain in equilibrium. In general, if a change in the lower-level structural components of a system alters the exercise of its higher-level capacity, the system is structurally cohesive. Systems that are interactively hierarchical are *interactively cohesive*, where a change in the organized interactions among lower-level components affects the higher-level capacity. Organisms are exemplary. A change in the outputs of the heart affects the circulatory system's capacity to deliver nutrients; a change in the efficacy of the circulatory system affects the organism's strength, stamina, and more. In general, if a change in the lower-level interactions of a system alters the exercise of its higher-level capacity, the system is interactively cohesive. And, of course, many systems are both structurally and interactively cohesive. Organisms, again, are exemplary.

4. Lycan (1987) argues that there is no absolute distinction between functional properties and systemic structures implementing those properties. Rather, the function/structure distinction is always relative to a given level of nature, in which case the attribution of functions and implementing structures is always relative to a given analysis of capacities at that level. It should be obvious that the theory of systemic functions, as presented by Cummins and as developed here, takes Lycan's advice to heart. This should be obvious in several ways, including the fact that Lycan cites Cummins (1975) approvingly—see footnote 4 in chapter 4 of Lycan's discussion. The relevant portions of that (1987) chapter are reprinted in Lycan (1990).

The world is rife with complex hierarchical systems. Animals such as ourselves are endowed with a host of high-level capacities—memory, vision, reasoning, speaking, singing, fighting, dancing, etc.—by virtue of complex organized effects of lower-level components. Lower-level components include various systems—muscular, skeletal, digestive, cardiovascular, neurological, and the like. Each of these systems, in turn, is composed of complex organized effects of still lower components, including various sorts of tissues, cells, molecules, etc. And the organized effects at each of these lower levels give rise to capacities at higher levels. There are, moreover, complex hierarchical systems other than organisms. A group of organisms may constitute a system with higher-level capacities instantiated by the organized effects of the organisms. Beehives, for example, exercise various capacities by virtue of the organized effects of member bees. And if the thesis of chapter 3 is true, populations that evolve or remain the same as a result of natural selection are composed of structural or interactive components that give rise to the capacity to change or to resist change.[5]

On my view, then, the attribution of systemic functions is warranted only to the extent that the system involved is hierarchical. Significantly, this restriction is justified on the grounds that the theory is designed specifically to apply to such systems. Unless the specified system consists of two or more levels of organization, where the capacities at one give rise to those at the next, the theory of systemic functions cannot be applied. The point is not that applying the theory to nonhierarchical objects produces a paucity of explanatory information—though that is true. The point, rather, is that the theory, given its basic structure, *cannot* be applied in such cases. There is explanatory failure, to be sure, but the source of the problem is a failure of application. After all, the central aim of the

5. I also believe that complex artifacts, such as computers and cameras, as well as simpler artifacts, such as books and coffee mugs, have systemic functions by virtue of their roles in various systems of production and consumption. These latter systems are formed from various social norms—various coordinated expectations we adopt and various sanctions we impose. As suggested in chapter 1, the centrality of intentions and conventions suggests that the functions of artifacts are importantly different from those of natural, nonengineered traits. That is why I reserve the topic of artifact functions for some other occasion.

theory of systemic functions is to enable us to break into systems composed of two or more levels of organized mechanisms. This aim is clear from Cummins's constraints, according to which a systemic function analysis is appropriate only when the higher-level capacity is produced by lower-level capacities that are less sophisticated and different in kind. But, of course, a capacity at one level is produced by simpler and different capacities at a lower level only when the system is hierarchically organized—only when higher-level capacities emerge out of structural and interactive capacities among lower-level components. Cummins makes the point in explanatory terms—in terms of whether or not the *analysis* is interesting and informative; but we can also make the point in ontological terms—in terms of whether or not the *system* has the requisite hierarchical structure. The upshot is that the theory, by virtue of the conditions it sets forth, applies only to systems that are hierarchically organized.

Restricting the theory in this way results in a proliferation of functions, or at least fails to prevent their proliferation, since even simple natural objects are composed of lower-level components (molecules, for example) that give rise to higher-level capacities. Some may take this as an objection, but I do not. A central aim of science is to discover how natural phenomena do what they do. The aim is to discover how natural systems work to produce the effects they produce. To say that systemic functions are ubiquitous is to say, at least in part, that hierarchically organized systems with higher-level capacities are ubiquitous. And, of course, that is true. An adequate theory of functions must be generous precisely because so much of nature consists of hierarchically structured systems.

This is not to say, however, that all component capacities are functional capacities. Some systemic capacities—perhaps most systemic capacities—are functional. But not all are. And a central burden on any theory of functions, as Enç and Adams (1992) emphasize, is to provide principled grounds for distinguishing between mere systemic capacities and systemic capacities that are functional. The theory of systemic functions discharges this burden by virtue of its internal structure—by virtue of applying only to hierarchically organized systems. The basic thought is that a mere capacity is one that contributes little or nothing to some more general capacity of the containing system. A mere capacity contributes little or nothing to the workings of the larger system. The heart's vibrating the sternum, for example, contributes nothing to any more-general capacity of the organism. A genuine systemic function, by contrast, is a capacity the effects of which contribute to the exercise of some more-general capacity of the larger system. The heart's pumping blood, as opposed to its vibrating the sternum, is illustrative—and the theory of systemic functions captures this difference. To apply the theory, we identify some higher-level capacity of the system and analyze it into the capacities of components at some lower level of organization. Crucially, our analyses require that we identify just those lower-level components the effects of which *in fact* contribute to the specified higher-level capacity. This is the cash value of the phrase "appropriately and adequately" in condition (ii) of the theory (chapter 2, section II). We identify at the lower level of organization just those capacities the effects of which in fact contribute structurally or interactively to some higher-level capacity; those systemic capacities thus are functional. Other capacities of the system that do not contribute in this way—"mere" capacities—are omitted from our systemic functional taxonomy.

The point bears repeating. Restricting the theory to hierarchically organized systems provides resources with which to distinguish the functional from the merely causal. A system is hierarchical to the extent its capacities at one level of organization are caused by the organized capacities operating at some lower level of organization. These latter capacities can be structural or interactive in character, but it must be the case that they, in concert, produce the higher-level capacity. In the course of analyzing such a system, we thus identify capacities at one level that contribute to the exercise of some capacity at a higher level. Capacities at lower levels that contribute nothing to the specified higher-level capacity are left out of our analysis. This enables us to say that, to the extent that our analysis of a system is correct, none of the identified capacities are mere capacities; they are all systemic capacity functions. Precisely this is why my view is not "eliminativist" in the sense described by Enç and Adams (1992).[6] It is not eliminativist because it accords a role to the attribution of functions distinct from the attribution of mere capacities. And precisely

6. See chapter 1 for discussion of eliminativism.

this is why restricting systemic functions to hierarchically organized systems ought to be judged a virtue of the theory—because it enables the theory to distinguish the functional from the nonfunctional on highly principled grounds.[7]

Restricting systemic functions in this way offers two additional virtues. Both are concerned with the practical utility of the theory and both are cited in the splendid work of Bechtel and Richardson (1993).[8] The first is that many natural phenomena, most notably organisms, are in fact hierarchically organized. Organisms are composed of several layers of organized mechanisms that give rise to the structures and capacities at higher levels. A theory designed to uncover the operations of such layered systems is a desirable theory. The second reason concerns the capacities and limitations we bring to the study of hierarchical systems. This is a large topic, of course, but two considerations are suggestive. One is accessibility. In systems that are hierarchical and interactively cohesive, often the efficacy of salient mechanisms is hidden from view. We cannot fix our attention or even our experimental methods directly upon them. Rather, we must prod or deprive the system in order to substantiate inferences concerning the mechanisms responsible for the higher-level capacity. Conceptualizing the system in terms of a hierarchy of levels facilitates inquiry into capacities at levels not directly accessible. Research in cognitive psychology illustrates the benefits of this approach. A further consideration is tractability. Systemic function analyses allow us to focus on the capacities and mechanisms at one level of organization while bracketing considerations of other levels. It also allows us to focus on a proper subset of the capacities at a given level. We can analyze the circulatory system into its components and usefully attribute systemic functions without simultaneously pursuing mechanisms at levels below the circulatory system

7. This discussion hints at the extent to which the attribution of systemic functions is shaped by the role such attributions play in the course of scientific inquiry. I address this topic in chapter 6.

8. Bechtel and Richardson (1993) are not in the business of developing or defending the theory of systemic functions. They are, however, interested in articulating two strategies of inquiry that we bring to the study of hierarchical systems. And these strategies overlap much of the theory of systemic functions. See chapter 6 for defense of this claim.

and also without pursuing all of the mechanisms at a single level. This is not to deny that there are contexts in which our investigatory ends require an analysis of more than one level at once. We may, for example, need to study the molecular structure of heart tissue to explain some feature of the heart's capacity to expand and contract and hence to explain the circulatory system's capacity to deliver nutrients. The point is simply that, when the variables involved overwhelm us, we can apply the theory of systemic functions by temporarily partitioning the relevant phenomena, thus providing some degree of tractability.

A further psychological consideration, drawn from Chase and Simon (1973) and discussed by Bechtel and Richardson (1993), concerns the way experts perceive and organize information in the relevant domain. Experts in a given domain recognize and remember domain-specific information on the basis of hierarchically derived patterns:

> Expert chess players, for example, are readily able to reconstruct board positions of games that they have observed for only a few seconds, whereas novices are able to locate only a few pieces. This difference disappears, however, when the board positions do not make strategic sense. Chase and Simon [1973] contend that the differences are due to the fact that experts recognize and remember patterns among pieces and treat these patterns as units. (Bechtel and Richardson 1993, 28)

Experts come to see and recall on the basis of higher-level patterns, ignoring the more complex lower-level interactions that give rise to these patterns. Abstracting from nuts and bolts and attending instead to high-level patterns is surely a commonplace among a wide range of experts—mathematicians, logicians, musicians, philosophers, architects, and more. This suggests that we come to inquiry with the capacity to conceptualize the systems we study at levels of organization that we find tractable. We abstract from systemic complexity by focusing upon high-level patterns of organization. We then respond to the system by reasoning about these patterns. And it seems plausible that constraints on time, memory, and computing capacity incline us this way. If this is correct—if we are so inclined—it is no wonder that we find the theory of systemic functions, applied to hierarchical systems, a fruitful strategy of inquiry.

Restricting the theory of systemic functions to hierarchical systems explains the power of Cummins's constraints. Cummins claims that a systemic analysis is interesting and informative only if the analyzed capacity is explicated in terms of analyzing capacities that are simpler and different in kind. I agree, but only because such considerations are symptomatic of underlying structural and interactive features of hierarchically organized systems. A higher-level systemic capacity is analyzable into simpler and substantially different lower-level capacities only if the system is hierarchically organized. The power of Cummins's constraints thus derives from the fact that the theory of systemic functions is designed to apply to systems in which higher-level capacities are produced by the structural and interactive cohesiveness among lower-level capacities.

The theory of systemic functions ought to be reformulated to reflect the requirement that the system be hierarchical. Where "A" refers to the analysis of system S into components, and where "C" refers to the systemic capacity we wish to explicate, item I has systemic function F if and only if:

(i*) I is capable of doing F,

(ii*) A appropriately and adequately accounts for S's capacity to C in terms of the organized structural or interactive capacities of components at some lower level of organization,

(iii*) I is among the lower-level components cited in A that structurally or interactively contribute to the exercise of C,

(iv*) A accounts for S's capacity to C, in part, by appealing to the capacity of I to F,

(v*) A specifies the physical mechanisms in S that instantiate the systemic capacities itemized.

Condition (ii*) requires that our analysis of the system "appropriately and adequately" account for the system's higher-level capacity in terms of lower-level capacities. As noted above, an analysis is appropriate only if it identifies just those capacities the effects of which in fact contribute to the specified higher-level capacity; extraneous lower-level capacities—mere capacities—must be left out. And an analysis is adequate only if it identifies all of the capacities involved in producing the specified higher-level capacity. Analyses that are appropriate and adequate in these ways represent the ideal toward which we strive. Of course, that ideal is not always met because condition (v*) is not always satisfied (more on this in chapter 6). More to the point of this chapter, and as we will see presently, this reformulation of the theory undermines the alleged counterexamples raised by Millikan and Matthen.

It might be objected that vagueness afflicts our concept of systemic hierarchy. Haugeland's fiber-optic cable may be hierarchical in the relevant way, but no doubt we can find cases in which our confidence is diminished. And if so, then my reformulation of the theory represents little progress—or so it might be claimed. Now, I agree that we likely can find cases that strain our concepts, but I do not agree that restricting the theory of systemic functions to hierarchical systems is of no utility. I take it for granted that the question whether or not a given system is hierarchical in the relevant way will be answered in the course of inquiry. Many cases are relatively clear. Organisms as complex as Welsh Corgis are clearly hierarchically organized. But when faced with borderline cases, we should turn to the results of inquiry for relevant evidence. If, in practice, we mistake a nonhierarchical system for a hierarchical one, our attempts to apply the theory to such a system are apt to bear little or nothing in the way of knowledge. Attempts to locate lower-level mechanisms instantiating higher-level capacities will fail; attempts to predict or otherwise control the system's behavior will fail; attempts to take the system apart and rebuild it will fail.[9] So, the extent to which our concept of hierarchical systems can be made fully explicit is of little importance. In general, we should be skeptical when faced with objections to a theory that rest so heavily on claims concerning conceptual intuitions. Caution is particularly prudent when the theory at issue concerns a cluster of concepts—functions, purposes, design, etc.—having a largely theological history. At any rate, and as I argue in chapter 6, the way to evaluate the theory of systemic functions is not by examining this or that concept, but

9. This is not to revert to Cummins's explanatory constraints. The point is that, in difficult cases, we should rely on the verdicts of the relevant but fallible experts. That seems to me our best bet. In the end, however, whether or not an object is hierarchically organized depends on the way it works to produce its capacities.

rather by assessing the extent to which inquiry into natural phenomena is facilitated by applying the theory. The aim of a theory of functions is to conceptualize functional properties in ways that fit the methods and postulations of our best natural sciences. We thus should be guided by the role or roles that function attributions play in the course of inquiry. If proceeding in this way forces upon us certain adjustments in our conceptual repertoire, that is all to the good.

Finally, the above reformulation of the theory is tentative. I leave open the possibility that additional conditions may be in order, as well as refinements on the conditions offered. Such additions and refinements, I believe, should be made on the basis of fruitful applications of the theory within actual scientific inquiry, rather than ruminations on the alleged content of our concepts. The aim of this reformulation is therefore modest. My aim is to push back the charge of promiscuity leveled against Cummins's version of the theory by one rather simple alteration to that theory. Let us consider, then, the extent to which my reformulation achieves this modest goal.

III The Alleged Counterexamples Revisited

Begin with the case of circulatory noise. As we saw, Cummins's constraints rule out the claim that the heart has the systemic function of making noise. Since the noise produced by the heart is neither simpler than nor different in kind from the noise produced by the circulatory system as a whole, the resulting systemic analysis is uninformative and thus we refrain from any such function attribution. While this response is plausible, I prefer to cast the issue in terms of properties of the specified system rather than properties of the resulting explanation.

The first thing to notice in this regard is that the circulatory system is not the relevant system, given the specified capacity. The capacity we wish to explain is some type of noise apparently emanating from certain parts of the circulatory system. Now, according to the theory of systemic functions, we analyze from the top down, beginning with the capacity we wish to explain and itemizing the lower-level components involved. Proceeding in this way, however, we quickly discover that not all components of the mammalian circulatory system produce noise. Capillaries make no appreciable noise in the process of delivering nutrients. Nor do critical chemical reactions within the blood. Nor the electrical signals that control the expansion and contraction of the heart muscle. Thus, these and other components within the circulatory system are not part of the system relevant to the production of the specified noise. They are part of the circulatory system, but not the "circulatory noise system." Our top-down analysis does not extend to them.

Which parts of the circulatory system are included in our analysis? It all depends upon the specific higher-level noise we are trying to understand. Cummins does not specify any particular type of noise, but suppose we want to understand the production of noises that cardiologists hear through a stethoscope. Suppose further that our top-down analysis reaches only as far as the noises produced by the motions of the heart and the whooshing of blood through arteries. Capillary noises, if any, are not included on the grounds that they cannot be detected with only the aid of a stethoscope. Similarly for other components belonging to the circulatory system. In consequence, our hypothetical "circulatory noise system" includes just these noises—the thumping of the heart and the whooshing of blood—and it accounts for these capacities by citing just three component parts—the heart, arteries, and blood. It should go without saying that these three components do not a circulatory system make.

This is an important point in assessing the force of the alleged counterexample. The example as Cummins describes it appears to embarrass the theory of systemic functions. It seems to force the theory to attribute to the heart the function of making a noise despite the fact that, within the circulatory system and relative to the systemic capacity of delivering nutrients, the heart has no such function. In fact, however, no such function attribution is forced upon the theory of systemic functions. Once we notice that the relevant system is not the circulatory system as a whole, but merely the heart, the arteries, and the blood, the charge of promiscuity dissolves. The relevant "system" under analysis is not the circulatory system but rather a small aggregate of noisemakers that also happens to be part of the much larger circulatory system. So, the theory of systemic functions does not warrant the attribution of the function of making noise in the context of analyzing the circulatory system. At worst, it warrants the attribution of making noise in the context of analyzing a quite different "system"—a system that has as one of its components a small aggregate of noise-making traits.

Is this an embarrassment to the theory? No, it is not. The relevant systemic capacity, we are supposing, is the production of noise detectable with a stethoscope. And the relevant "system" consists of the heart, the arteries, and the blood. The crucial issue, then, is whether or not this purported system is hierarchically organized. The question is whether or not the production of noise by these traits contributes to the exercise of some higher-level systemic capacity. If not, then the theory as I have defended it does not warrant the attribution of any systemic function, since these traits constitute no more than a mere aggregate. If, however, the production of noise contributes to the exercise of some higher capacity, then the attribution of a systemic function goes through. Exactly what this higher capacity might be is far from obvious. But assuming that there is a higher capacity to which the production of these noises contributes, then, crucially, the heart has the systemic function of making noise *relative to that system.* It does not follow, and in fact it is not the case, that the heart also has the systemic function of making noise relative to any other system, including the circulatory system.

Some may find this unsatisfying. It may be tempting to insist that the heart cannot have the function of making noise no matter what the larger system to which it belongs. The heart, it may be objected, has the purpose of pumping blood but not the purpose of making noise, even if noise-making is useful in this or that respect. This is to advert to Millikan's distinction between a trait genuinely having a given function and a trait merely functioning as something. And that, as I have indicated, is a distinction we ought to relinquish. The argument for this claim is developed in subsequent chapters.

Consider now Millikan's water cycle. Recall that Cummins's constraints defeat this alleged counterexample. If the systemic capacity of the water cycle is to distribute moisture on or near the surface of the earth, then we cannot attribute to clouds the systemic function of producing precipitation, because producing precipitation is neither simpler than nor different in kind from distributing moisture on the earth. If the systemic capacity is to contribute to vegetable growth, then the water cycle is not the appropriate system, since by itself it is incapable of growing anything. A much larger system is required. The water cycle may be one component of this larger system and within this system the cycle may have the systemic function of distributing moisture. But if so, then once again we cannot attribute the function of producing precipitation to the clouds, since that is neither simpler nor different in kind from the function of the cycle as a whole. While this response to Millikan is persuasive, the root problem, once again, concerns the alleged system involved.

The problem is that the water cycle is not a hierarchical system. It is a mere cycle. It is composed of a sequence of events, including (1) evaporation, (2) cloud formation, and (3) precipitation (the distribution of moisture). Millikan's mistake is to take one of these steps—the distribution of moisture—and treat it as a higher-level capacity of the others. The distribution of moisture is no more a higher-level capacity of the cycle than evaporation or cloud production. Indeed, no one of these steps is constituted from the organized effects of all or some of the others. Each causes the next but does not constitute it, does not instantiate it. There is no hierarchical relationship between steps (1) and (2) and step (3). This should become clear to us in the course of constructing a top-down analysis of the distribution of moisture. But if it is not clear there, it becomes obvious in our bottom-up search for the right sorts of systemic mechanisms.[10] In consequence, the theory of systemic functions cannot be applied to the water cycle; it is not the right sort of system. Of course, this conclusion is compatible with the claim that the steps in the water cycle, in the context of some larger hierarchical system, may possess systemic functions. Like the noises produced by the heart, there may well be a natural system within which rain clouds contribute salient effects to the exercise of some higher-level capacity. But the water cycle, taken by itself (as Millikan does), is a mere cycle, not a hierarchical system.

It is instructive to compare the water cycle to two other cycles. The menstrual cycle consists of a sequence of causally related events. Hor-

10. As I say above (note 2), vegetation admittedly plays a role in the water cycle today by producing oxygen and absorbing carbon dioxide. But to insist that these capacities of vegetation be included in our inquiry is to insist that we expand the relevant system well beyond the water cycle. And expanding the relevant system will, of course, affect the attribution of systemic functions.

monal changes cause the formation of a lining in the uterus and the dropping of ova. If fertilization does not occur, the lining breaks down and is sloughed off. Then the sequence recurs. This sequence, however, unlike the water cycle, occurs in the context of the reproductive system, and that makes all the difference. Each step in the sequence contributes to the higher-level systemic capacity to produce offspring. The point, then, is that each step in the sequence has a systemic function not because of its placement within the regular sequence of events, but rather because of the structural or interactive role it plays in the systemic capacity to conceive. There are systemic functions not because there is a cycle, but because there is a hierarchically organized system. Contrast the menstrual cycle with the cycle of the seasons. This too consists of a sequence of events, though the sequence does not occur in the context of a larger system. There is no higher-level systemic capacity to which the occurrence of autumn or winter contributes. Rather, each step in the sequence is no more than a byproduct of the orbit of the earth around the sun. The orderly procession of seasons forms a cycle of sorts, but not one in which the various steps are structurally or interactively organized. Each step has a common cause, but there is no higher-level systemic capacity to which any step contributes. The attribution of systemic functions to the occurrence of autumn or winter is thus not warranted.

The water cycle, like the seasons, is not part of a hierarchical system. Or, more modestly, as Millikan describes the case, there is no hierarchical system to which the water cycle belongs. Millikan seems to think that the theory of systemic functions is refuted on the basis of considering the water cycle on its own. But unless the steps that form the cycle contribute to the exercise of some higher-level systemic capacity, the theory of systemic functions does not properly apply. Unless the water cycle can be shown to be closer to menstruation and further from the seasons, Millikan's alleged counterexample does not engage the theory of systemic functions.

My response to Matthen's tusks is simpler than my response to Millikan's water cycle. Matthen asserts that it is possible to give an analysis according to which the systemic function of the unwieldy narwhal tusk is the reduction of mobility. Unfortunately he provides no such analysis. Matthen also asserts that the function of the unwieldy tusk is to increase sexual attractiveness and not to reduce mobility, but he provides no argument for this claim. Perhaps he takes this point to be intuitively obvious. My response, however, is that Matthen's way of formulating his claim has a certain degree of rhetorical force but no argumentative force. For it is quite easy to give a systemic analysis within which the function of the unwieldy tusk is to reduce mobility; it is easy to place the effects of the tusk in a hierarchically organized system. And the resulting function attributions are plausible on their face.

If we assume, as Matthen seems to assume, that the function of the unwieldy tusk is to increase sexual attractiveness, then presumably the systemic capacity we wish to explain is the relative increase in reproduction among whales with especially large tusks. Cast in terms of the view defended in the previous chapter, large tusks have the selected systemic function of increasing sexual attractiveness—a systemic function that arises in the context of a system affected by selection. Relative to the organism's capacity to reproduce at a certain rate and relative to the population's capacity to evolve, the systemic function of the large tusk concerns increased sexual attractiveness. But, of course, we can be concerned with organismic capacities other than the capacity to reproduce at some rate and we can be concerned with capacities of the population other than the capacity to evolve or remain in equilibrium due to selection. It is easy to imagine some other organismic or population capacity within which the function of the tusk differs.

Consider, for example, the capacity of the large-tusked whale to fight. Suppose that, as larger tusks emerged in the population, there occurred a decrease in combat efficiency among large-tusked whales. As tusk size increased, mobility and speed decreased, reducing the number of battles won. At the organismic level, we explain this loss of fighting efficiency among large-tusked whales by pointing to the effects of their various components, especially the unwieldy tusk. We explain, that is, a change in the exercise of a certain capacity—the capacity to win fights—by appeal to the capacity of a specific phenotype—the large tusk—combined with various other phenotypes. This is an appropriate application of the theory of systemic functions. Our analysis applies at the population level too. We might notice, for example, that the frequency of large-tusked whales increased after the larger tusk first appeared in the population, but then leveled off after a few generations, and we might want to explain this leveling off. As I describe in chapter 3, we analyze a population into the relevant structural units—whales with large tusks and those with less cumbersome tusks—and attribute to these units the appropriate structural systemic functions. We then iterate our strategy and break into the structural units and analyze their structures and interactions. We thus break into the morphology of the large-tusked whale and identify the salient effect of the large tusk. And we discover, let us suppose, that large tusks reduced mobility and thereby reduced the capacity to win fights.

There is nothing untoward in this sort of analysis. If something seems amiss, the reason, I speculate, is because two quite different systemic analyses are run together. This conflation, I suggest, is what gives Matthen's alleged counterexample its rhetorical force. If our systemic analysis is focused on the way selection has affected the organism or the population, then the function of the large tusk probably is to increase sexual attractiveness—though relevant historical evidence must be adduced (more on this in chapter 5). Relative to this analysis, it is indeed unintuitive to attribute to the tusk the function of reducing mobility. But set that analysis aside and focus instead on explaining how the increase in large-tusked whales leveled off after the initial increase. Focus, that is, on the capacity of the population to reach a kind of equilibrium. Within this analysis, we can attribute to the large tusk the task of reducing mobility. This is not to attribute a "selected" systemic function to the tusks since, by hypothesis, the reduction in mobility was not selectively efficacious. We may even assume that reduced mobility was selected against during the relevant stretch of time.[11] There are, after all, many systemic capacities we wish to explain and understand besides those that result in selective success. So long as we keep these sorts of systemic analyses separate, the intuition that seems to support Matthen's alleged counterexample disappears.

11. As suggested in chapter 3, we can explain the capacity of the population to evolve or remain in equilibrium by appeal to selection and also by appeal to nonselective forces such as drift. This example illustrates the fact that we can explain the capacity of a population to achieve or remain in equilibrium by appeal to effects of traits that are selectively disadvantageous. In this case, reduced mobility contributed to the selective *in*efficaciousness of large tusks and, in consequence, the population's capacity to reach a kind of equilibrium.

This reply to Matthen, like my reply to Millikan, may fail to satisfy some advocates of selected functions. Some may complain that it fails to account for the very real sense we have that some natural traits are properly functional while others are merely useful. As indicated, I disagree that any natural traits are endowed with the property of being "properly" functional. I concede that some natural traits strike us as more functional than others. But this, I suggest, should be understood as a fact about our psychology, not the natural traits we are trying to understand. In chapters 5 and 6, I speculate on the causes of this sort of reaction in us. Explaining why some things strike us as more functional than others is one step in defending my claim that there are no such things as natural purposes or functions that are "proper" to nonengineered, natural traits.

IV Systemic Functions or Self-Perpetuation

Before closing this chapter, I want to consider the view defended in Enç (1979).[12] Enç urges that we turn away from the analysis of the concept of functions—a tradition distilled in Wright (1973, 1976) and Woodfield (1976)—and instead examine the role of function attributions in the course of scientific inquiry. In the course of his discussion, he accomplishes three things. He (1) offers necessary conditions for the attribution of functions that overlap the substance of the theory of systemic functions, (2) criticizes these conditions on the grounds that they fall prey to the promiscuity objection, and (3) adds a further condition intended to dissolve the objection, concluding that the conditions offered are necessary and jointly sufficient for the attribution of functions. In response, I wish to make two points. I shall argue that the theory of systemic functions, when constrained by the requirement of hierarchy, dissolves Enç's version of the promiscuity objection. I then shall argue that my solution to this problem is preferable to the one endorsed by Enç.

12. Enç (1979) represents a very early step away from the generic form of conceptual analysis as practiced by Wright and others, toward a consideration of the role of function attributions within well-developed forms of inquiry. For similar steps, see Ayala (1970), Wimsatt (1972), and Brandon (1981). Enç's essay also represents an early wedding of key elements from the theory of systemic functions and the theory of selected functions. It is a highly original paper.

Simplified, the core conditions necessary for the attribution of a function, according to Enç, are as follows: Trait T in organism O has function F only if

(a) Under certain conditions, T undergoes certain movements or changes of state,
(b) Under certain further conditions, T's moving or changing in this way is causally sufficient for the performance of task F,
(c) The performance of F by T contributes to some property or capacity of O,
(d) In those cases where the performance of F is part of the identity conditions of trait T, T is the only kind of thing that under normal conditions is capable of F-ing in all Os.[13]

I wish to set aside discussion of conditions (a) and (d), except to say that cases such as the evolution of melanism in the peppered moth persuade me that (a) is not a necessary condition of function acquisition. But for present purposes the real interest of these conditions is the extent to which (b) and (c) overlap the theory of systemic functions. It is clear that, on Enç's view, the attribution of a function to a systemic component is accomplished relative to some larger systemic capacity to which the functional component contributes. The emphasis, crucially, is on functions as effects that contribute to more general systemic capacities.

After defending these conditions, however, Enç raises the promiscuity objection in the following manner:

> the planet Neptune occupies a series of positions in the solar system, and this series of positions are [sic] causally sufficient for the perturbations in the orbit of Uranus; furthermore, without these perturbations the solar system would not exhibit the properties it does, e.g., the conjunction of Uranus and Earth would occur at different times. However, it is clear that the function of Neptune is not to cause the deviation of Uranus from its perfectly elliptical path. (Enç 1979, 361–62)

The charge is that, because Neptune has the capacity to perturb the orbit of Uranus and thus affect the conjunction of Uranus and Earth, we are

13. I have recast and simplified Enç's conditions, leaving out several important details that are not to the point of this discussion. I refer the reader to p. 359ff of his (1979) discussion.

thus forced to attribute to Neptune the function of perturbing the orbit of Uranus. This example, it seems, satisfies conditions (b) and (c) of Enç's formulation. The orbit of Neptune causes perturbations in the orbit of Uranus and this, in turn, has the systemic effect of altering the point at which Uranus and Earth are nearest one another in their orbit. Were it not for the perturbations caused by Neptune, the conjunction of Uranus with Earth would occur differently. The orbit of Neptune thus appears to possess the function of perturbing Uranus, despite the very strong intuition that Neptune has no such function. And this is to say that conditions (b) and (c) are objectionably promiscuous in the function attributions they warrant.

I agree that the case of Neptune poses a challenge to conditions (b) and (c) of Enç's view and I will consider his response presently. Meanwhile, given the overlap between Enç's conditions and the theory of systemic functions, it is natural to worry that the case of Neptune poses a problem for my view as well. But, in fact, this is not the case. The strategy employed in the previous section is relevant here. It is relevant insofar as the solar system is not a hierarchical system. Or, more modestly, the solar system is not hierarchical in the way required by the Neptune case. The deviation in the orbit of Uranus is not a more-general capacity of the solar system. Nor is the conjunction of Uranus and Earth. Moreover, the orbit of Neptune is not a lower-level capacity of the solar system. It is, I believe, clear on its face that neither phenomenon is higher or lower, systemically speaking, than the other. In the context of the solar system taken as a whole, it is implausible that some feature of the orbit of one body is a higher-level capacity while the orbit of any other body is a lower-level capacity. There is a causal relationship, to be sure, between Neptune's orbit and Uranus's deviated orbit, but similarly there is a causal relationship between evaporation and precipitation in the water cycle. On my view, the systems to which the theory of systemic functions applies must be hierarchically organized. Such systems must exercise higher-level capacities as a product of the organized effects of lower-level capacities. It is hard to see that the case of Neptune satisfies this requirement.

Suppose we reverse the capacities described by Enç. No doubt the orbit of Uranus, including its deviation, has some causal effect on the orbital pattern of Neptune. After all, if the gravitational pull of Neptune affects the motion of Uranus, it is likely that the gravitational pull of Uranus exerts some effect on the motion of Neptune. We thus might hold that the deviation in the orbit of Uranus has some systemic function insofar as it exerts some influence on the orbit of Neptune. We might hold, that is, that the former is a lower-level capacity of the solar system while the latter is a higher-level capacity. But this, too, is implausible on its face. The solar system is like the water cycle and the cycle of the seasons; there is a sequence of events that tends to recur, but no hierarchical system. Insofar as we can reverse the alleged functional relationships between the phenomena described by Enç, we should conclude that these phenomena are not hierarchically related as required by my version of the theory of systemic functions.

The alternative, I suppose, is to allow that the solar system is hierarchical in some minimal sense—perhaps in the way that the composition of salt is conceptualized in hierarchical terms. Of course, we should allow that this is plausible only to the extent that astronomers find it theoretically fruitful to conceptualize the solar system in such terms. But if they do, we should accept the claim—in which case the attribution of some systemic function may go through. Much depends on the higher-level capacity of the system we are trying to understand. It may turn out that there is no capacity relative to which the systemic function of the orbit of Neptune is to perturb the conjunction of Uranus and Earth, but we must leave open the possibility that some such capacity exists. The attribution of functions, as I argue in chapter 6, should be motivated by the fruits of our best-developed scientific theories, even when they conflict with our pre-theoretical intuitions. So, if some such capacity is found, then we must reject Enç's intuition that the orbit of Neptune is functionless—just as we reject Matthen's intuition that large tusks cannot have the function of reducing mobility. The situation, then, is that either the solar system is not a hierarchical system, in which case the theory of systemic functions does not apply, or it is hierarchical with respect to some more-general capacity, in which case the intuition to which Enç appeals is defeated by the discovery of some theoretically significant systemic function. The case of Neptune, therefore, is no objection.

Enç's own solution to the charge does not appeal to hierarchy. Instead he draws on the central insight of Wright (1973, 1976)—an insight preserved in the theory of selected functions—that function attributions explain why the functional trait has persisted or proliferated. In addition to conditions (a)–(d), Enç further requires that

(e) Part of the explanation of why trait T normally undergoes certain movements or changes in state lies in the fact that such movements or changes result in the performance of task F.

Unlike Wright and advocates of selected functions, Enç does not require an explanation that appeals exclusively to some process of selection. Instead, he requires that, in attributing to T the function of doing F, we thereby assert that part of the reason for T's continued operations (movements, changes) is because such operations result in the performance of F. The central idea is that causal effects are also functional effects only when they contribute to the perpetuation of the token trait or the trait type. Effects that fail to contribute to the perpetuation of the trait are "mere capacities." Consider, then, a token heart in a token organism. It continues to exist in the organism—continues to expand and contract—because such expansions and contractions result in the pumping of blood. The movements or changes that result in the pumping of blood contribute to the maintenance of life, which in turn contributes to the continued movements or changes in the token heart. The token heart thus satisfies condition (e). Consider, too, a type of heart within a population. It persists across generations—and thus continues to pump blood—because such pumpings contribute to differential reproduction, which in turn contributes to the perpetuation of tokens of this type that pump blood. So, once again, (e) is satisfied.

Appeal to (e) blocks the Neptune case on the plausible assumption that Neptune's perturbing the orbit of Uranus contributes nothing to the perpetuation of Neptune's orbit. That is, the alleged functional effect—perturbing Uranus—is not part of the explanation for why Neptune's orbit persists. Were the perturbations to cease, Neptune's orbit would continue on as before, and this shows that the perturbations contribute nothing to the perpetuation of Neptune and its orbit. So Neptune's perturbing the orbit, while a clear causal effect, is not a functional effect. It is precisely this kind of connection between functional effect and the perpetuation of the functional object that Enç wishes to capture in (e).

Now, I agree that a theory of functions is adequate only if it successfully distinguishes functional effects from mere causal effects. And I believe that (e) is, at first glance, a compelling way to make the relevant distinction. But upon reflection, I do not think that (e) is necessary for functions. On Enç's view, the difference between the functional and the merely causal comes to this: Functions are those systemic effects that contribute to the perpetuation of the functional trait. Or, as Schlosser (1998) puts it, functions are systemic effects that contribute to the "re-production"—the recurrence—of the functional token or the functional type.[14] I suggest we reject this attempt to distinguish the functional from the merely causal. Two considerations are relevant. First, and as noted in section II, the theory of systemic functions, when restricted to hierarchically organized systems, already draws this distinction. In hierarchical systems, capacities at one level are produced by organized capacities operating at some lower level. A systemic function analysis of a hierarchical system thus aims to identify just those lower-level capacities that in fact contribute to the exercise of some higher-level capacity. Capacities that contribute nothing to the exercise of the specified higher-level capacity are left out of the analysis. Thus we can mark the distinction between effects that are functional and those that are merely causal from within the theory of systemic functions and we can do so without inheriting the burdens of the theory of selected functions.[15]

Second, Enç holds that functional effects are those that contribute to their own perpetuation, while mere causal effects are ones that fail to so contribute. Similarly, Schlosser holds that functions are effects that contribute to their own re-production. But this, I believe, is not the case. Consider yet again Haugeland's fiber-optic cable. The capacity of each individual fiber within a token cable hardly contributes to the persistence of that token. Such fibers are structural features merely; their efficacy does not ensure their own persistence. Of course, were we to peel off individual fibers one at a time, at some point the cable would lose the

14. See chapter 2 for a synopsis of Schlosser's view.
15. These burdens include the charge that the theory is redundant (chapter 3), that it faces significant epistemic and methodological challenges (chapter 5), and that it fails in its self-appointed task of explaining how malfunctions are possible (chapter 7).

capacity to transmit an image, but the issue is whether or not the exercise of the capacities of each fiber somehow ensures the persistence of the cable. They do not. The same point holds, moreover, for various natural traits. Consider the genes that contribute to larval development in the fruit fly *Drosophila melongaster.* Specific genes turn on at quite precise points during development and then turn off at quite precise points, never to turn on again. The efficacy of such genes within a token organism hardly contributes to their future efficacy within that token.

Of course the capacity of each token fiber or each token gene does contribute to the perpetuation of the relevant type. It is because each fiber transmits a quantity of light that we continue to produce these cables. That is plausible enough, but it is a mistake to think that this is always the case. Each fiber within a token cable has a systemic capacity and this would be so even if we produced only a single cable and then refused or found ourselves unable to perpetuate the type. Or imagine an artifact in which some component has the express function of ensuring that future tokens of itself (the type) are not produced and that it (the token) does not endure. Consider, for example, a computer program designed to create a new generation of software programs. Suppose we want this first program to ensure that programs produced in the second generation differ from it in some way. And suppose the only way to accomplish this is to include in the first generation a device with some systemic function along with instructions that this device is not to be reproduced. The function of this device, we may suppose, is required only in the initial generation and its persistence would either be idle or destructive in subsequent generations, so we include instructions for its elimination immediately after it has done its job. Such examples may be odd or rare, appearing only in episodes of *Mission Impossible.* But such devices surely are intelligible. And it seems clear that they would possess rather important systemic functions, despite the absence of Ençean perpetuation or Schlosserean re-production.

Nor are such cases biologically impossible. It is at least intelligible that a genetic mutation results in the production of some phenotypic trait—an enzyme, for example—that serves some systemic function but nevertheless contributes nothing to its own perpetuation. This may occur when the resulting trait contributes to some organismic capacity but the mutated gene is not heritable. Such a trait, like the device programmed into the first generation of software, can have a clear, identifiable systemic function even though nothing it does contributes to its own re-production or perpetuation. Alternatively, and as I argued in chapter 3, the theory of systemic functions applies to systems and systemic capacities that do not involve selection, in which case there exists a broad range of systemic functions that do not involve the perpetuation of the functional item. The unwieldy tusk of the narwhal discussed above may have the systemic function of reducing mobility without contributing to the perpetuation of that clumsy trait. Indeed, it may have this systemic function even while contributing to its own elimination. Thus, it seems clear that the justified attribution of functions does not require effects that are self-perpetuating.

I concede that many functions, as a matter of fact, are attributed to items that, in the course of producing their functional effects, contribute to their own perpetuation (tokens or types). This is a consequence of two facts. One is that biological components belong to living systems and living systems tend to reproduce their own kinds. The second is that artifactual components belong to systems that we produce in order to satisfy recurring desires, in which case we tend to produce such systems with regularity. So, self-perpetuation or some form of re-production is a feature ubiquitous to functional traits. But ubiquitous features need not be constitutive. And if the examples just described are intelligible, as I think they are, then re-production is not constitutive of functions despite being ubiquitous.

If this is correct, then the entire tradition that originates in Wright (1973) and matures in Enç (1979) is in error. It may be tempting to explicate functions not merely in terms of contributions to systemic capacities, but more narrowly in terms of contributions to systemic capacities that bear directly on systemic perpetuation. But if I am right, it is a mistake to narrow our account of functions in this way. Functions are effects that contribute to the exercise of more-general systemic capacities; this is so even when the more-general capacities contribute nothing to systemic perpetuation. The capacity of some systems to re-produce themselves is, to be sure, a capacity we wish to understand and control. But it is not the only kind of systemic capacity worth studying. And that is why the tradition instituted by Wright—a tradition that persists in the historical approach generally—ought to be set aside in favor of the theory of systemic functions.

V Conclusion

The promiscuity objection, I conclude, is compelling only when the applicability of the theory of systemic functions is determined by nothing more than explanatory considerations—only when the appropriate sorts of "systems" and "systemic capacities" are fixed entirely by the "interests" of the investigators. But when the theory is restricted to systems that are hierarchically organized, the promiscuity objection falls away. By contrast, the theory of selected functions attempts to explain our inclination to see some things as genuinely functional by attributing genuine norms of performance to certain natural objects. This theory takes for granted that our inclination to see some things as genuinely functional is best explained by postulating "proper" functions; our inclination is explained by appeal to norms of performance that attach to natural traits. And this, I believe, is at odds with the aim of giving an account of functions that fits a naturalistic view of the world. I turn now to consider the difficulties involved in preserving, from within our naturalistic worldview, belief in functions that are "proper."

5

Naturalizing Functions: Evidence, Methodology, and Ontology

. . . truth can never be opposed to truth.
—John Herschel 1831

The historical and systemic approaches to functions aim to naturalize our understanding of functions. They aim to provide an account that fits the larger view emerging from our natural sciences. Advocates of selected functions claim success in this regard, emphasizing that selected functions are the products of natural selection and that the theory of evolution by natural selection is a highly confirmed, well-developed scientific theory. A close connection to evolutionary theory, it is claimed, secures enviable naturalistic credentials for the theory of selected functions. The thesis of this chapter, however, is that this connection is not nearly so cozy as advocates would have us believe and, in fact, that the naturalistic credentials of the theory are open to serious doubt. The dubious credentials of this theory contrast starkly with those of the theory of systemic functions, which I explicate in chapter 6. This contrast provides further reason for rejecting the theory of selected functions, construed as a separate and autonomous theory, and advocating the theory of systemic functions instead.

A theory of functions is naturalistic only if it is open to empirical or theoretical test. It must be possible to discover that the theory is in error on empirical grounds or that it conflicts with the methods or postulations of well-developed scientific theories. This I take for granted. This is not to eschew a priori considerations. If our account leads to contradiction, we have grounds for rejecting it; if it leads to conceptual incoherence, we have grounds for revising or perhaps abandoning it. But a theory can be free from a priori defects and nevertheless fail to be naturalistic. There must be conditions concerning the actual world—conditions to which we have empirical and theoretical access—against which we can test the claims of the theory. Naturalism requires that we evaluate our theories against the world we inhabit.

There are two levels at which to test a theory of functions. One is the level of specific function attributions. If the conditions specified within the theory render the attribution of functions immune to test or difficult to substantiate, we have reason to doubt the naturalistic standing of the theory. Particular attributions must be open to compelling empirical or theoretical tests specified by the theory. The second level concerns the orientation of the theory as a whole, especially its ontology and methods of inquiry. If the theory commits us to an ontology or methodology at odds with those in well-developed and highly confirmed scientific theories, then once again we have reason to doubt the theory's naturalistic credentials. As I now shall argue, the theory of selected functions performs poorly at both levels.

I Confirming the Attribution of Selected Functions

Begin with the attribution of a selected function to some specific type of trait or some specific token. To assert that trait T has selected function F is to assert that tokens of T persisted or proliferated because the performance of F among ancestral tokens was selectively efficacious. The function attribution is true only if the history of the trait fits the bill—only if ancestral tokens performed F and were selectively successful as a consequence. The attribution is confirmed, therefore, only if the relevant historical evidence is available and favorable.

"Availability" of evidence should not be understood narrowly. Evolutionary biologists often speculate about selective history in the face of little or no direct evidence. Indirect evidence, while not ideal, is acceptable. Of course, some forms of indirect evidence are better than others. Kingsolver and Koehl's (1985) explanation of the evolution of insect wings involves compelling indirect evidence, drawing heavily upon empirical tests performed on physical models of fossilized flies (see chapter 3, section III). The evolution of melanism in the peppered moth (chapter 2, section I) and the evolution of metal tolerance in certain grasses (Brandon 1990) are even more persuasive, based upon observations of selection in progress. These and other examples may inspire optimism about the evidential credentials of selected function attributions; they give the appearance that the theory of selected functions is indeed naturalistic. But, in fact, no such optimism is warranted. Taken as a separate, autonomous theory, the theory of selected functions is intended to apply quite generally within the biological realm (and the artifactual realm too, at least for some advocates). But the availability of evidence in a few cases does not a general theory naturalize.

At least three difficulties threaten the naturalistic credentials of selected function attributions. One concerns the availability of even indirect historical evidence. Often the required historical evidence has deceased and rotted away without a trace. Kingsolver and Koehl were fortunate to have fossils from which to construct physical models, but speculations concerning the functions of most traits are not nearly so secure. Behavioral and psychological traits, in particular, tend to leave nothing at all or nothing but archeological remains from which it is difficult to draw conclusions with confidence. And, of course, environments evolve, exacerbating the difficulty in guessing the ancestral structure and selective efficacy of certain traits. These and other difficulties, common to the study of evolutionary history, force us to settle for untested (and perhaps practically untestable) speculations in our how-possibly explanations. And this, in turn, makes the attribution of selected functions correspondingly weak. This is an obvious point that should trouble advocates of selected functions: The historical evidence for the attributions of such functions is often weak, sometimes entirely absent.

The evidential weakness of most attributions of selected functions does not deter advocates, however. It is often argued that, since selection is the only natural process capable of producing adaptive complexity among natural traits, we are thereby justified in attributing selected functions in the absence of historical evidence. So long as there are marks of complexity, or at least marks of functional efficiency, the attribution is warranted. This argument is an explicit part of the methods of so-called "evolutionary" psychology[1] and may be implicit in the theory of selected functions. But the argument is invalid. Two considerations are relevant. First, even if we accept that complexity cannot arise without the aid of selection, it does not follow that every component within a complex system is responsible for the selective success of the system. It does not follow that every systemic component is selected for. Some may have been merely selected of—pulled along by virtue of their connection to other components that were selected for. And this may occur even when the components selected of now play important roles within the economy of the larger system. (The example of skull sutures discussed below is illustrative.) Given this possibility, the mere existence of complexity does not justify the attribution of selected functions to every efficacious trait within the system. What is required is evidence concerning the actual selective history of the system; otherwise we have no grounds for claiming of any specific trait that it was selected for.

Second, to assert that trait T has selected function F is not merely to assert that ancestral tokens of T were selected for. It is to claim, more fully, that ancestral tokens were selected *specifically* for their performance of F. To establish this, we need evidence that (i) ancestral tokens had the capacity to F, (ii) their doing F was relatively selectively efficacious (relative to organisms that did not have T), and (iii) the selective efficacy of their doing F was not overridden by the efficacy with which other traits performed other tasks. Acquiring these sorts of evidence, however, is no easy matter. Two considerations make this clear. First, (ii) requires evidence that there was variation in the population with respect to trait T and (iii) requires information concerning the effects of other traits within the system, information that rules out the possibility that some other selective or nonselective process neutralized the selective efficacy with which tokens of T performed F. All this makes for a rather tall order of evidence. Second, while some historical hypotheses can be ruled out on the basis of comparative phylogenetic studies—as Griffiths (1996) and others have suggested—the utility of those studies in attributing selected functions is limited. The relevant sorts of studies compare the historical trajectories of traits across related species. Thus, we learn that low birth

1. See, for example, Cosmides and Tooby (1987) and (1992). See also Pinker and Bloom (1992).

weight in bears is not related to hibernation, since low birth weight evolved before hibernation and continues to occur in related species that do not hibernate (McKitrick 1993). This is valuable information, to be sure, but it is far too coarse-grained to be of much use in the attribution of selected functions. After all, selected functions are usually highly specific causal effects, in which case the information required in (ii) and (iii) is correspondingly specific. We especially need evidence that the effects of T were not overridden by the effects of other organismic traits; we need evidence that selective success is the result of T and not some other feature of the organism. Comparative studies are not likely to cast light into corners such as these. It is one thing to trace the trajectory of a given trait in two or more branches of a phylogenetic tree; it is quite another to show that the efficacy of T in one branch (or set of branches) was not neutralized by the selective efficacy of some other trait in the organism's economy.

These considerations show that the mere fact that T is part of a complex system does not suffice for the attribution of selected functions. There is no dodging the difficult task of adducing historical evidence.[2] And this means that the vast majority of selected function attributions are best construed as historical hypotheses awaiting the discovery of evidence. We thus should be wary of the confidence with which advocates ascribe selected functions.

Considerations of historical evidence are problematic in a second way. The problem is not the paucity of historical evidence, but its evident irrelevance. The problem stems from the simple fact that practicing scientists often attribute functions with no concern for the evolutionary history of the traits involved. Scientists often appear to make important discoveries about substantive functional properties with no consideration of selective or evolutionary histories. The point is not that these scientists fail to unearth the evidential details described in the previous paragraph; the point rather is that they offer highly compelling attributions of functions with nary a word about evolution or history. Consider the challenge raised by Nagel (1977).[3] He presses the following objection against Wright's

2. I press both these points against the methods of so-called "evolutionary" psychology in Davies (1996) and again in Davies (1999).

3. See Boorse (1976) for a similar argument.

(1976) theory of etiological functions, but the point generalizes to the theory of selected functions:

> [Wright's] analysis requires us to say that F is a function of an item *i* if and only if the item had been selected in some way to be present in the organism just because F is an effect of *i*'s presence. It therefore follows that F can be *asserted* to be a function of *i*, if and only if it is *known* (or there are good reasons for *believing*) both that F is an effect of *i* and that *i* had been selected to be present in the organism just because F is an effect of the item *i*. In fact, however, biologists commonly do state (often on the basis of experimental findings) that a function of some item *i* in organism S is F, but without knowing or believing that one *causal determinant* of *i*'s presence in S is that F is an effect of *i*. (Nagel 1977, 284)

Important functions often are attributed in the absence of justified beliefs concerning the selective history of the relevant trait. Nagel cites William Harvey's discovery of the function of the heart, a discovery made long before the formulation of the theory of evolution by natural selection. Harvey justifies his claims concerning the function of the heart on the basis of his experimental work, not historical speculations, and certainly not an undiscovered scientific theory. He justifies his claims on the basis of overwhelming evidence concerning how the circulatory system in fact works. Speculations on how the system evolved are completely absent. Now, we surely do not want to hold that Harvey was unjustified in assigning a function to the heart. In that case, however, we must conclude that the historical evidence—if any—concerning the heart was quite beside the point.

Advocates of selected functions have attempted to rebut Nagel's challenge. Millikan (1989), for example, insists that her version of the theory of selected functions is not an analysis of the concept "function" but rather a *theoretical definition*. This is supposed to answer Nagel's challenge. According to Millikan, the assertion that water is H_2O is a theoretical definition; so too is the assertion that gold is the element with atomic number 79. A theoretical definition, in this sense, is an account not of a concept but of stuff in the world and its underlying structure, and the definition is true or it is false quite apart from the structure of our concepts. Likewise, the theory of "proper functions" (Millikan's term for selected functions) is a theoretical definition. It is an account of the origins and nature of functions, and it is true or it is false quite apart from the concept employed by us or by life scientists. So whatever Harvey had in mind concerning the nature of functions is quite irrelevant to the truth of the theory of selected functions. Unless Harvey had in mind the theory of selected functions—and presumably he did not—then, on Millikan's view, Harvey's notion of functions was in error. This is so even if the extension of the concept of functions employed by Harvey happens to overlap function attributions warranted by the theory of selected functions. Any such overlap is nothing more than a happy coincidence. It does not suffice to show that Harvey's nonselectionist conception of functions is correct.

This seems a weak response to Nagel's challenge. Nagel is challenging the theory of selected functions by citing a case in which a bona fide functional property was discovered without any reference to or beliefs about natural selection. Millikan's response is to claim that, insofar as Harvey did not have the relevant theoretical definition of functions in mind—insofar as he did not have Millikan's theory in mind—his account of functions is mistaken. But unless Millikan's theoretical definition has been confirmed on independent grounds, and unless it has been shown that selective success is in fact constitutive of functions, we ought to approach her response to Harvey with caution. Unfortunately, Millikan's definition of functions is far from confirmed.[4] Indeed, I believe that it is either straightforwardly false or at least insulated from any reasonable form of test. To see this, notice first that, when Lavosier hypothesized that water is H_2O, there were reasonably clear methods with which to test his hypothesis. And the methods available today are clearer still. We need only collect samples of the stuff we typically find in our rivers and lakes and examine its microstructure in the laboratory. Similar considerations apply to the hypothesis that the atomic number of gold is 79. But in what laboratory are we supposed to test the theory of selected functions? And exactly what is the stuff we are supposed to sample? The analogy between a definition of water or of gold and a definition of functions is strained. We may allow that testing a definition of functions involves considerations of overall theoretical coherence or explanatory

4. For criticism of Millikan's definition independent of the considerations offered here, see Neander (1991), Davies (1994), and Preston (1998).

power, rather than laboratory procedures, but the question remains: How do we test a theoretical definition of functions for coherence and power? Remember that, by Millikan's own lights, even expert use of the concept of functions (Harvey's, for example) is entirely irrelevant to the truth of her theory. What, then, is relevant?

It turns out that, on Millikan's view, the relevant issue is the selective history—indeed, the selected function—of the relevant theoretical term. A term can be given a theoretical definition only if it has a selected function; its having a selected function is integral to its being theoretically defined. And the selected function of such a term is, roughly, to designate the stuff in the world that accounts for the selective success of the term itself. More specifically, the selected function of a referring term is to designate whatever in the world accounts for the selective success of those language users who interpreted or "consumed" past tokenings of the term.[5] Suppose, then, that the term "water" has been perpetuated in the language because it referred, often enough, to H_2O, enabling us to interact successfully with H_2O, causing us to perpetuate the term. In that case, "water" has the selected function of designating H_2O precisely because reference to H_2O explains the selective success of the term. This theoretical definition of water thus is confirmed just in case there is evidence concerning the selective success of the term; historical facts settle the functional status of the term. Likewise for the assertion that functions are selected functions. We need historical evidence concerning the selective success of the term "function." If this term has been perpetuated because, often enough, it referred to the selectively efficacious effects of the specified trait, and if the historical record shows that this is so, then Millikan's theoretical definition is confirmed.

This view is odd in at least two ways. First, it appears to march blindly into Nagel's challenge. Suppose we accept Millikan's view. Suppose we agree that, if the perpetuation of "function" is a consequence of the fact

5. I am abstracting from Millikan's account just those elements required to substantiate my criticisms. I am leaving out, for example, the claim that a theoretical definition is intended to express the "sense" of the relevant term, where "sense" refers to whatever in the world the term is "supposed to" correspond to. For the ingenious full story, see Millikan (1984), especially chapters 6 and 8. For further hints of the larger view, see notes 7 and 8 below.

that past tokenings of the term in some way corresponded to the selective efficacy of the specified trait, then the theory of selected functions is confirmed. On this standard, the fact that Harvey applied the term "function" with obvious felicity *without* referring to the selective efficacy of the heart suggests that the historical evidence required by Millikan's view is not forthcoming. The success of Harvey's nonselectionist concept of functions is part of the historical record and, on its face, it conflicts with the historically based theoretical definition that Millikan has put forth. Indeed, it is reasonable to speculate that the perpetuation of Harvey's use of the term "function" is due to the fact that past tokenings corresponded to the internal operations of the circulatory system—to Harvey's experimental findings. In addition to Harvey, consider the eighteenth- and nineteenth-century English deists who espoused the argument from design. Talk of "design," "purpose," and "function" proliferated among those writers and worked its way into the early thought of Charles Darwin.[6] Needless to say, these terms did not proliferate by virtue of referring to the natural selective history of the relevant traits; selection was rather far from the mind of Paley (1802) and his ilk. And it would be mere table pounding to insist that Paley's use of these terms was somehow "supposed to" correspond to the selective success of relevant functional traits.[7] At any rate, it is up to Millikan to provide historical evidence that the historical perpetuation of "function" among English deists and others is a consequence of some sort of correspondence with the selective success of relevant traits. Since no such evidence has been proffered, Millikan cannot dismiss Nagel's challenge out of hand.

6. See Browne (1995) for a splendid account of Darwin's thought up to and throughout his voyage on the Beagle.

7. As mentioned, Millikan holds that referring terms have a "sense." Having a sense, she says, is being supposed to correspond to something; having a sense is having the selected function of corresponding to something even in cases where no correspondence obtains. So perhaps Millikan could dismiss the historical cases that conflict with her theory—Harvey and the English deists—and insist that, on her theory of reference, the intentions or thoughts of the relevant language users are not decisive. This may appear plausible on the assumption that a term has the selected function of corresponding to some object even when competent speakers fail to intend any such correspondence. But this line of thought drives us toward the problem described in the next paragraph, namely, that Millikan's theoretical definition of functions is objectionably circular.

But suppose we waive this first problem. Suppose we discover that the term "function" has indeed persisted because often enough it designated the selective success of the relevant functional trait. Nevertheless, the proposed definition of functions is unacceptably circular: Confirmation of the proposed definition *presupposes* that the definition has already been established. We are directed to look to the selective history of the term "function" as confirming the definition, where the definition itself asserts that functions are constituted from selective history. We are told, that is, to look to the *selected function* of "function" as confirmation for the theory of selected functions.[8] But, of course, the selective history of "function" is relevant and compelling only if we have already accepted the truth of the theoretical definition—only if we have already accepted that the theory of selected functions is true. If we have yet to pass judgment on the definition, then the selective history of the term hardly counts as confirmation of that definition. So even if we discover that "function" has indeed been perpetuated because it referred to the selective efficacy of traits to which it was applied, that tells us nothing about the evidential credentials of Millikan's definition of functions. It is hard to see, therefore, how Millikan's version of the theory of selected functions provides a suitably naturalistic account of functions.[9]

Neander (1991) replies to Nagel's challenge differently. She allows that scientific inquiry involves a dialectic between the structure of our concepts—the way we conceptualize the relevant phenomena apart from any specific theory—and the theoretical definitions we then formulate and test. Our concepts help us delimit the phenomena for which we seek a

8. It is hard to construe Millikan otherwise: "I do have a theory about what theoretical definitions are, a theory about how the theoretical definition of 'theoretical definition' should go. Unfortunately, this theory rests upon a theory of meaning that rests in turn on the notion 'proper [selected] function,' the very notion under scrutiny" (Millikan 1989, 291). Indeed! And in a footnote to this passage, she says: ". . . according to my theoretical definition of 'theoretical definition,' what a theoretical definition analyzes is (Millikanian) sense" (291, note 2). As indicated above (note 7), according to Millikan's theory, the "sense" of a referring term is determined by its *selected function* with respect to correspondence.

9. So far as I can see, there is no argument in Millikan (1984 or 1993) for the claim that her theoretical definition of functions is open to genuine test.

theoretical account and our theoretical definitions often force us to adjust our concepts to better fit the structure of the world. Progress in inquiry involves this sort of back and forth between concepts and definitions. Neither our concepts nor our theoretical definitions are static; both evolve as inquiry proceeds. Neander's response to Nagel, then, is quite simple:

> it is unproblematic if Harvey's notion of a "proper function," before the Darwinian Revolution, was different from the closely related notion used by biologists today, after the Darwinian Revolution. Scientific notions are not static. Harvey obviously did not have natural selection in mind when he proclaimed the function of the heart, but that does not show that modern biologists do not have it in mind. (Neander 1991, 176)

Neander assumes that the aim of a theory of functions is to give an account of the (explicit or implicit) conditions under which modern biologists attribute functions (more on this in section III). She grants that Harvey may not have had in mind the selective history of the heart but claims that our concept of functions has evolved since the inception of the theory of evolution by natural selection and, furthermore, that biologists today do have selection in mind.

It is easy to agree with Neander that our lay concepts and our theoretical definitions are dynamically related. It thus is plausible that the concept of function employed by Harvey may have differed from the one used by modern biologists. But it is also easy to restate Nagel's challenge in a way that evades Neander's response. A change of examples suffices. Instead of Harvey, consider the Revolutionary Himself. In chapter 3 (section III), I quote an excerpt from the *Origin* in which Darwin refers to two organs in a single type of fish, both of which have the function of providing the organism with oxygen. He suggests that a structural modification in one organ that alters its functional efficacy can result in the organ's perfection or in its obliteration. The claim is that, as functions change due to modifications in structure, so selection can act and so evolution may occur. Natural selection thus acts on preexisting functions. Indeed, Darwin's reference to a trait's function appears driven by considerations of the roles played within the organism's internal economy; they appear driven, as Nagel puts it, by the results of experimental findings, not historical ones. And I take it that Neander's reply in the case of Harvey is not available, or at least not plausible, in the case of Darwin.

Advocates of selected functions may be unmoved by the example of the fish. They may insist that the antecedent functions to which Darwin refers are nothing but selected functions acquired in earlier selective regimes. But this response is dubious at best; we are left to wonder what, if anything, would count as disconfirming the theory of selected functions. But we can bypass this worry by switching examples, for Darwin clearly believed that there were important natural functions that are *not* the products of selection. In a section of the *Origin* entitled "Organs of little apparent importance," he describes traits that have functions discoverable on experimental grounds, but traits for which we have historical evidence that they were *not* selected for their functional effects. The point is not simply a paucity of evidence in favor of selected functions, but rather positive historical evidence that weighs *against* the attribution of selected functions to traits that are clearly functional in a systemic sense of that term. Consider, for example, the skull sutures in young mammals:

> The sutures in the skulls of young mammals have been advanced as a beautiful adaptation for aiding parturition, and no doubt they facilitate, or may be indispensable for this act; but as sutures occur in the skulls of young birds and reptiles, which have only to escape a broken egg, we may infer that this structure has arisen from the laws of growth, and has been taken advantage of in the parturition of the higher animals. (Darwin 1859, 197)

The conjecture is that the sutures serve a highly useful function during birth among mammals despite the fact that sutured skulls have not competed against nonsutured skulls. The historical evidence, in sum, is that there has not been selection for sutures. The point, again, is not that there is no evidence for the selectionist hypothesis; the point is that there is evidence for a *non*selectionist hypothesis. Of course, the conjecture may be false, but so long as we restrict ourselves to available evidence, the attribution of selected functions is unwarranted. And the fact that this conjecture comes from Darwin and not a pre-Darwinian casts doubt upon the strength of Neander's reply to Nagel's challenge. So the challenge stands: There are important scientific discoveries—Harvey on the heart—involving the attribution of highly confirmed functions that in no way depend upon the availability of evidence concerning evolutionary history. And there are compelling function attributions—Darwin on skull sutures—despite evidence that the relevant trait has not been selected for.

The relevance of the historical record to the actual attribution of functions is not nearly so tight as the theory of selected functions seems to require.

The third difficulty threatening the naturalistic standing of selected function attributions concerns the identification of organismic traits responsible for selective success. The problem here is not with the historical record. The problem involves practical difficulties in identifying the organismic causes of selection even in contemporary populations. Consider the challenge raised by Amundson and Lauder. When studying a population, often it is difficult to determine which trait is causally responsible for the organism's response to the pressures of the selective regime. This is so, at any rate, for organisms in which several traits are linked by genetic or developmental mechanisms. In such cases, several traits will evolve in response to some selective pressure and "it is extremely difficult to separate the individual trait that is responding to selection from those that are exhibiting a correlated response" (1994, 461). In fact, it often is easier to identity the generic effect that is being selected for—the effect answering to the selective demand—rather than the specific trait responsible for the effect.

To illustrate, suppose we have evidence that there has been selection for flight duration in a population of insects. Organisms unable to remain aloft for extended periods of time mate less frequently and thus are selected against. We may, in such a case, have strong evidence that selection favored a particular effect—flight duration—but be unable to specify any particular traits responsible for this effect. There will be positive correlations between several organismic traits and the increase in flight duration:

> It is almost certain, in fact, that many aspects of muscle physiology, nervous system activity, flight muscle enzyme concentrations and kinetics, and numerous other physiological features would show correlated change in mean values with the increase in flight duration. In addition, body length and mass are likely to show positive correlations, as are wing length, area, and traits that have no obvious functional relevance to flight performance (such as leg length). If we cannot identify the causal relationships among these correlated variables to single out the one that was selected for, we will be unable to assign a trait X to the [selected] function already identified. (Amundson and Lauder 1994, 462)

Cast in terms of the theory of systemic functions, we may put the point this way: We may identify the systemic functions of various traits within the organism, relative to the larger systemic capacity of, say, survival or relative reproduction. But in some cases we will be hard pressed to say which trait or which proper subset of traits was directly responsible for the organism's selective success. We should not assume that every systemic function within an organism is selectively efficacious;[10] we cannot know, *a priori,* that every systemic function within an organism in fact contributed to selective success.[11] We need to ascertain, for each particular case, the range of traits targeted by the demands of the selective regime. And often it is practically difficult, if not impossible, to identify those involved. It is agreed that selection does indeed affect some organismic traits. The question is, Which ones? In some cases the answer may be easy to discern, but in many cases the correlations adduced as evidence do not distinguish between traits that are selectively efficacious and those that are merely correlated with selective success. Thus, the difficulties involved are quite real. And lurking just behind this practical difficulty is a rather well-known theoretical difficulty, to which I now turn.

II Confirmation and the Causes of Selection

The theory of evolution by natural selection suffers from what Brandon (1982) calls the "levels of selection" problem—what I prefer to call the "causes of selection" problem. When a type of entity evolves as a consequence of natural selection, which properties of that entity are causally responsible? When, for example, an organism evolves due to selection, which properties of the organism are causally responsible for its selective success—phenotypic, genotypic, some combination of these two, or something else?[12] And on what grounds do we settle on one answer over

10. This is why I say, in chapter 3 (section III), that applying the theory of systemic functions to systems affected by selection does not alleviate the evidential burdens involved.

11. For one thing, we need an account of the causes of selection in order to individuate traits that are among the targets of selection and thus among those potentially selected for. More on this in the next section.

12. According to developmental systems theory (Oyama 1985; Griffiths and Gray 1994), phenotypes emerge out of a range of messy causal factors ranging

the others? These questions bear directly on the theory of selected functions. Advocates of selected functions assume that their theory applies to a wide range of organismic traits—genes, cells, tissues, organs, subsystems, psychological capacities, behaviors, and more. This assumption is justified, however, only if a defensible account of the organismic causes of selection establishes that traits at all these levels of organization qualify as targets of selection. This is because a trait can be selected for and thus endowed with a selected function only if it can be targeted and sorted by demands of the selective regime. Unfortunately, and as I argue below, no advocate of selected functions has produced an account of the causes of selection with sufficient breadth to justify the attribution of selected functions to the various levels of organismic organization. The embarrassing result is that the attribution of selected functions often is unwarranted on grounds internal to the theory. And this should give us pause about the naturalistic status of selected functions.

Before defending this claim, I want to distinguish the causes of selection problem from the so-called "units of selection" problem. The latter problem refers to a cluster of questions, as Lloyd (1992) makes clear, and the evolution of altruistic behavior raises one such question in an especially pointed way. Altruistic behaviors decrease the fitness of the agent while increasing the fitness of others. The central theoretical task is to explain how and why such behaviors have evolved. After all, as Darwin describes it, evolution by natural selection involves a struggle for existence, and the naive view is that individual organisms are driven entirely by considerations of self-interest, in which case altruistic behaviors would always

from genes and maternal cytoplasmic elements to a host of "environmental" elements such as exposure to language, play behavior, etc. On this view, the "causes of selection" question is that much harder to answer, since candidate causes include not only phenotypes and genotypes but also the tangled developmental interactions that give rise to phenotypes. I would reject the suggestion, however, that such interactions render the causes of selection question unintelligible. Inquiry into the causes of selection can be challenging, to be sure, but so too is inquiry into the cause of most any event, requiring us to differentiate relevant from irrelevant historical factors. Moreover, the importance of the causes of selection question to the theory of selected functions can be made without countenancing the developmental systems perspective.

be selected against.[13] Attempts to explain the evolution of such behaviors in terms of natural selection must demonstrate that, in addition to sorting among individual organisms within a population, selection also sorts among groups of organisms within a larger collective of groups. Such attempts must show that selective forces discriminate among groups of organisms in ways that redistribute such groups in succeeding generations, parallel to the way selection discriminates among organisms within a population. And this is to say that an adequate account of altruistic behavior must articulate and defend the claim that selective pressures discriminate among at least two different kinds of units—individual organisms (within a single population) and groups of organisms (within a more global population of groups). To defend this thesis is to defend a particular view of the units of selection.

Sober and Wilson (1998) offer an exciting account of the evolution of altruistic behavior in which they adopt a pluralistic view of the units of selection. They argue, in particular, that selection sorts among a variety of entities—genes within a single organism, individual organisms within a population, but also groups of organisms within a more global population of groups. It is important to see, however, that the focus of their inquiry is distinct from the question concerning the causes of selection. Sober and Wilson argue that, so long as there are groups of organisms that reproduce differentially in response to pressures within a common selective environment, selection discriminates among groups of organisms. Cast in general terms, the thesis is that selection discriminates among different kinds of biological entities. By contrast, an account of the causes of selection is *not* committed to any particular view concerning the units of selection. We can be neutral on whether or not selection ever discriminates, for example, among groups of organisms. We need only assume that selection discriminates among *some* biological entities. Now, I take it for granted that selection discriminates among organisms within populations; I take it for granted that individual organisms are among the entities of selection. That is why I raise the problem concerning the causes of selection by asking, Which properties of an organism—genotypic, phenotypic, both, or some other—are responsible for its selective

13. There is, of course, the famous passage from *The Descent of Man*. See Darwin (1871), 166.

success? But a parallel question applies to genes, to groups of organisms, and to any other type of entity sorted by selection. So the question concerning the causes of selection is not, What kinds of entities can evolve via selection? but rather, For any type of entity that can evolve via selection, which among its various traits are causally responsible for its selective trajectory? This is to inquire into the selective efficacy of entities affected by selection.

Consider the population of the peppered moth *Biston betularia*. It evolved from mostly light-colored to mostly dark-colored moths during the second half of the nineteenth century near Manchester, England. After studying the selective regime and the capacities of the moths, Kettlewell (1973) discovered that dark coloration in the moth's wings was causally salient—it camouflaged moths from predatory birds. But we also know that pigmentation for dark coloration is controlled by alleles at a single chromosomal location. So, which properties of the moth performed the selective work—dark coloration, the alleles that code for dark coloration, or both? And why accept one answer over any of the others? This is not to ask whether moths qualify as units of selection. The question presupposes that they do. The question, rather, is, Which properties of the moths are among the causes of selection? And the question I wish to consider is, To what extent do answers to the question concerning the causes of selection cohere with the theory of selected functions?

It is remarkable that, among the many advocates of selected functions, virtually none has offered an answer to this question. Brandon (1982, 1990) is an exception. In addition to defending the theory of selected functions, he defends a view of the causes of selection.[14] He agrees with Mayr (1963) and Gould (1980) that phenotypes, not genotypes, are the causes of selection, but his reasons are novel. Borrowing from Salmon (1971), Brandon claims that, because phenotypes *screen off* genotypes with respect to selection, only phenotypes can be causes of selection. Screening off requires two probabilistic relations among events or causal interactions. Event B screens off event A from outcome C if and only if

$$P(C/A) = P(C/A\&B) \neq P(C/B) \qquad \text{(S)}$$

14. See Brandon (1981, 1990) for defense of selected functions. I discuss Brandon's version of the theory further in chapter 7.

That is, A screens off B from C just in case the probability of C given A alone is equal to the probability of C given A and B, and the probability of C given B alone is less than the probability of C given A and B. When this relation holds, A renders B statistically irrelevant with respect to outcome C. Applied to selection, the effects of phenotype p screen off the effects of genotype g with respect to relative reproductive success r if and only if

$$P(r/p) = P(r/p\&g) \neq P(r/g) \qquad \text{(SS)}$$

Thus the probability of superior reproductive success given the effects of the phenotype equals the probability of success given the effects of the phenotype and the genotype. The probability of success given the effects of the genotype, however, is less than the probability of success given the effects of the genotype and phenotype. Brandon's claim is that, when these relations hold, the phenotype renders the genotype statistically and causally irrelevant to selection. He also claims that these relations hold generally, or at least in most cases for most organisms, in which case phenotypes and not genotypes are the causes of selection.[15]

One way to defend this use of screening off is by way of thought experiments. Imagine a case where we alter some genotype without altering any phenotype that causally engages the demands of the selective environment. Doing so has no affect on organismic reproductive success. Additionally imagine a case where we alter some phenotype without altering any genotype. Castration vividly illustrates this second claim, as Brandon notes. Castrating an antelope eliminates its reproductive success but does not alter its genotypic endowment. The truth of both claims supports the equality and the inequality in statement (SS).

The considerations that favor Brandon's account of the causes of selection are compelling, though hardly immune to critical challenge.[16] But suppose for the sake of argument that his account of the causes of selection is true. Even so, it mixes poorly with his endorsement of selected

15. It is irrelevant whether or not Brandon is committed to the strong claim that phenotypes always screen off genotypes or the moderate claim that this relation typically holds. The criticism raised below applies to the moderate as well as the strong claim.

16. See Kane and Richardson (1990) and Sober and Wilson (1994).

functions.[17] Notice first that, if selection selects only (or mostly) at the level of the phenotype and not the genotype, there can be no (or few) genetic traits with selected functions. Selected functions accrue only to those traits causally responsible for their own selective success; they accrue only to traits causally responsible for their own self-perpetuation. Thus, to use Mayr's, Gould's, and Brandon's metaphor, since selection "sees" only phenotypes and not genotypes, genetic traits cannot be selected for. I take it, however, that genes and genotypes do possess important functions—functions that advocates of selected functions wish to account for. And while the theory that I endorse—the theory of systemic functions—warrants the attribution of functions to such traits, the theory of selected functions, given Brandon's view of the causes of selection, fails to warrant any such attributions. This should give us pause.

The problem is worse still. Brandon's view assumes that there is a reasonably clear distinction between genotypic interactions and phenotypic interactions within an organism. Without this distinction, it is difficult to see how Brandon's use of the screening off relation can hold. By parity of reasoning, then, we should be able to establish a parallel distinction between levels of organization among the phenotypes of complex organisms. We should be able to divide into complex phenotypes and discern distinct levels of organization. Consider, for example, the running speed of an antelope. This phenotype enables the organism to flee predators and thus reproduce another season. Just as Brandon assumes that phenotypes are the orchestrated effects of various genotypes, so too, for purposes of argument, we can assume that speed is the orchestrated effect of several lower-level phenotypes, including lung capacity, muscle strength, blood chemistry, etc. So, if screening off holds between phenotypes and genotypes with respect to differential reproduction, then it also holds between phenotypes such as speed and lower-level phenotypes such as lungs, muscles, etc. with respect to differential reproduction. If so, then screening off holds between levels of phenotypes with respect to selection.

17. Brandon has since amended his view and now accepts a version of the combination approach described in chapter 2. See Brandon (forthcoming). But the point raised above holds for his amended view as well as his earlier view, since the combination approach maintains the theory of selected functions as an integral part.

The upshot is that, on Brandon's view, while selection may select at the level of speed, it cannot select at the level of lungs, muscles, blood chemistry, etc. We thus are barred from attributing selected functions to lungs, muscle strength, etc. The scope of the theory of functions, given this account of the causes of selection, is unacceptably narrow.

I am not claiming that speed in fact screens off lung capacity and muscle strength with respect to reproduction. The claim, rather, is that this is a plausible hypothesis *if* we have already accepted Brandon's claim that phenotypes screen off genotypes. After all, the considerations that support his claim concerning phenotypes and genotypes also support my suggestion concerning levels of phenotypes. He asks us to imagine a change in some genotype that does not alter any phenotype. We likewise may imagine a change in some feature of the antelope's lungs or muscles that does not affect its speed. This is easy enough to imagine, especially if we are allowed to compensate for a loss of efficiency in one system by increasing efficiency elsewhere. You and I may be equivalent with respect to speed as a consequence of distinct ensembles of lower-level phenotypic capacities; your strong muscles and mediocre lungs may match my strong lungs and mediocre muscle strength. We can tinker with the various lower-level phenotypes without altering the higher-level trait. Brandon also asks us to imagine a change in phenotype that affects reproduction but does not involve any change among genotypes. Castration is his example. Of course castration will alter the antelope's reproductive potential without altering the efficacy of its lungs, muscles, etc. The same is true of other sorts of damage. Breaking the animal's legs or destroying its ability to hear will diminish its running speed and reproductive potential without altering its muscles or lungs. So the considerations that support Brandon's view also support the claim that some phenotypes screen off others with respect to selection.

I have not demonstrated that speed in fact screens off lungs and all the rest. Doing so may be a difficult task. But if we accept that screening off holds between phenotypes and genotypes, then we should also accept that some phenotypes screen off others, at least among organisms with sufficiently complex phenotypes. These other phenotypes, in consequence, are barred from possession of selected functions, in which case the scope of the theory is objectionably narrow. What advocates of selected functions need is an account of the causes of selection that preserves the broad scope of the theory of selected functions, that does not block the attribution of functions to genotypes and lower-level phenotypes. Until such a solution has been offered, we should not accept that the theory of selected functions is somehow grounded in the theory of evolution by natural selection. Nor should we accept the associated claim that the theory of selected functions naturalizes the attribution of functions.

It might be replied that, in focusing on Brandon's account of the causes of selection, I have overlooked an alternative account that holds promise for the attribution of selected functions to all levels of biological organization. The pluralism espoused by Sterelny and Kitcher (1988), it might be suggested, holds that organismic traits at various levels can be seen as causes of selection. The view is that, in some forms of inquiry, it may be fruitful to construe selection as affecting only the phenotype, while in others it may be fruitful to construe selection as targeting genotypes or even alleles. If this view is correct, then perhaps it is true that traits at these various levels of organization can be selected for. At the very least, it appears that traits at various levels can be construed as the targets of selection and hence as candidates for selected functions.[18] If so, then pointing out conflicts between Brandon's screening off and his endorsement of selected functions is no objection to the attribution of selected functions in general. That conflict may merely be reason to reject Brandon's monistic view of the causes of selection. In fact, however, the pluralism offered by Sterelny and Kitcher, no less than Brandon's

18. Sterelny and Kitcher's view is instrumentalist. Phenotypes may be construed as the causes of selection, but genotypes and even alleles may be construed similarly. Such pluralism, if coupled with the theory of selected functions, makes the attribution of functions similarly instrumentalist. From the point of view of one explanation, a given phenotype might be seen as having a selected function, but from some other point of view, that same phenotype might be seen as having some other selected function or none at all. This should make us doubt the extent to which selected functions derive naturalistic comfort from Sterelny and Kitcher's pluralism, for the theory of selected functions is a realist, not an instrumentalist, theory. I take it for granted that selected functions are supposed to exist insofar as selection in fact sorts among varying traits. If there is no fact of the matter concerning which traits are targets of selection, then, according to the theory of selected functions, there are no selected functions.

monism, raises doubts about the justification with which selected functions are attributed.

Sterelny and Kitcher are concerned to show that the arguments in the literature that favor one level of organization—the phenotypic over the genotypic, for example—do not work. The criticisms they offer motivate their pluralistic position. Their assessment of Brandon's view is illustrative. Brandon maintains that phenotypes screen off genotypes because (1) changes in phenotypes (castration, for example) with no attending changes in genotypes can affect reproductive success, while (2) changes in genotypes, when unaccompanied by phenotypic changes, cannot affect reproduction. Sterelny and Kitcher focus on (1). They agree that it is possible to change some phenotype affecting reproductive success without altering any alleles.[19] But this, they insist, in no way supports the contention that selection "sees" only the phenotype. The problem is that a phenotypic change affecting the organism's reproductive potential constitutes a change in the selective environment of the relevant alleles. Castration prevents the organism from reproducing and thereby prevents the reproduction of the organism's alleles; castration introduces an impediment into the selective environment within which the animal's alleles work to perpetuate themselves. Sterelny and Kitcher put the point this way:

> Champions of the gene's eye view will maintain that tampering with the phenotype reverses the typical effect of an allele by changing the environment. For these cases involve modification of the allelic environment and give rise to new selection processes in which allelic properties currently in favor prove detrimental. (Sterelny and Kitcher 1988, 353)

Brandon's appeal to screening off is compelling only if applied to phenotypes and alleles within a fixed selective environment. This is a point that Brandon has vigorously emphasized.[20] A change in the selective environment entails a change in selective pressures, in which case the probabilistic relations between phenotypes and genotypes in statement (SS) must be reassessed. The complaint raised by Sterelny and Kitcher, then, is that the

19. They cast their discussion in terms of alleles, rather than genotypes, to include within their pluralist embrace Dawkins's (1976) gene's-eye view of the causes of selection.

20. See chapter 2 of Brandon (1990).

thought experiments to which Brandon appeals involve an illicit change in the selective environment of the relevant alleles. Castration alters the organismic selective environment, to be sure, but it also alters the allelic selective environment. Once this change is made explicit, it becomes clear that screening off cannot hold. Indeed, if Sterelny and Kitcher are right, then *any* phenotypic change that alters the reproductive potential of the organism's alleles will count as a change in the allelic selective environment, in which case it is *impossible* for phenotypes to screen off genotypes with respect to selection. They put their conclusion this way:

> genic selectionists should propose that the probability of an allele's leaving *n* copies of itself should be understood relative to the total allelic environment, and that the specification of the total environment ensures that there is no screening off of allelic properties by phenotypic properties. (1988, 354)

This is a clever response to Brandon's account of the causes of selection, but it nevertheless mixes poorly with the theory of selected functions. Suppose we accept the pluralist's view of the causes of selection. In that case, our attributions of selected functions must be relativized to the sorts of environments described by Sterelny and Kitcher. Consider, for example, the alleles that code for bone tissue in an antelope. We may suppose that a broken leg qualifies as a phenotypic change that alters the animal's reproductive potential and thus alters the potential of its alleles. Before the leg break, the alleles operate in one selective environment; afterwards they operate in a new one. Since selection is environment relative, the acquisition of selected functions is likewise environment relative, and this means that a broken leg should force us to reassess and revise all of our function attributions after the break occurs. In the first environment, before the break, the antelope's alleles have certain selected functions. But after the break, the alleles are suddenly cast into a different selective environment, in which case we must suspend our previous function attributions and wait for selection to occur and sort among the alleles in this new environment. During that period, we must, I suppose, refrain from attributing any selected functions to the alleles, and after the period ends, we must determine anew the functions of these alleles. Indeed *any* phenotypic change affecting reproduction—broken limbs, loss of sight, loss of hearing, etc.—would force us to revise *all* of our prior function attributions in this way.

But this, I submit, flies in the face of biological practice, not to mention common sense. No one, including practicing biologists, would take seriously the suggestion that, upon breaking its leg, the functions of the antelope's alleles must be suspended and reassessed. Nor, moreover, is there any compelling reason to revise our concept of functions to fit the pluralist view of the causes of selection. We should require powerful arguments in support of pluralism to motivate revisions as radical as that, and so far the case for pluralism does not have that sort of clout. (By contrast, the attribution of systemic functions to the antelope's alleles is not affected by a broken leg.) This is not to criticize the pluralist view; indeed, pluralism may be the way to go.[21] This is to say only that Sterelny and Kitcher's pluralism meshes with the theory of selected functions no better than Brandon's monism.

It might be thought, finally, that it is unfair to saddle the theory of selected functions with the causes of selection controversy. After all, biologists and philosophers of biology have yet to resolve the question, so why expect advocates of selected functions to do so? This, however, is an admission of guilt, not a demonstration of innocence. Advocates of selected functions appeal to the theory of evolution by natural selection. It is *their* claim that evolutionary theory contains the resources with which to explicate our concept of functions. It turns out, however, that the question concerning the causes of selection poses a serious problem for this approach to functions; indeed, it poses a potential problem for anyone wishing to adopt the theory of evolution by natural selection to such ends. It is no good trying to dismiss or minimize the problem by pointing out that it is a problem at large within evolutionary theory. It is precisely because the theory of evolution by natural selection is problematic in this way that advocates of selected functions should reflect on the wisdom of trying to explicate functions in terms of that theory.

I conclude, therefore, that until the causes of selection problem has been resolved—and resolved in a way that justifies the attribution of selected functions at several levels of biological organization—the attribution of selected functions generally is unjustified. It is unjustified on

21. But see Sober (1990) for searching criticisms of Sterelny and Kitcher's pluralism.

grounds internal to the theory itself. This lack of warrant provides further reason to doubt the naturalistic credentials of the theory of selected functions.

III Confirming the Theory of Selected Functions

The considerations offered in the previous sections raise questions about the evidence for and theoretical coherence of the attribution of selected functions. These questions provide grounds for doubt about the naturalistic standing of such attributions. Troubling as they are, however, they are less worrisome than the problems raised by the methods and postulations of the theory taken as a whole. These problems, I shall argue, cast serious doubt upon the alleged naturalistic status of selected functions. I begin with doubts about the methods of the theory.

The theory of selected functions is heir to Wright's (1973, 1976) seminal work. Wright proposed that the question, "What is the function of item T?" is contextually equivalent to the question, "Why is item T here?" The suggestion is that the attribution of a function is essentially explanatory, explaining why the functional item exists in the specified system or the specified setting. The core of Wright's view is that an item has a function so long as it exists as a consequence of some form of selection. And it is Wright's appeal to a rather generic notion of selection that makes his view so prima facie compelling. In the case of artifacts, the relevant form of selection is deliberation among intentional agents. In the case of nonengineered biological traits, the relevant form is natural selection. The concept of selection involved in evolutionary theory, on Wright's view, is derivative from the more basic notion of conscious or deliberative selection. Wright thus assumes the existence of a quite general concept of selection that includes within its extension both intentional deliberation and natural selection. On this view, such selective explanations of why a trait exists also tell us what current tokens of the trait are for.

Advocates of selected functions preserve certain elements of Wright's view but reject others. They agree that the attribution of a selected function is equivalent to an explanation of the functional trait's existence—its persistence or proliferation in the population. The theory of selected functions is typically cast in such explanatory terms.[22] But advocates disagree about the scope of the concept "selection." Wright's notion is broad, embracing conscious and natural selection. By contrast, advocates of selected functions typically appeal solely to natural selection or simply reverse Wright's ranking and take natural selection to be fundamental. Conscious or subconscious selection is taken to be a specific instance of the processes involved in natural selection, if it is countenanced at all.[23] The motivation for making natural selection basic is, I take it, to tie the theory of functions directly to a well-developed, highly confirmed scientific theory, one that is independently plausible. It is also to suggest that one's view of functions somehow shares in the naturalistic standing of evolutionary theory.

Several theorists are explicit in their claim that the theory of selected functions is grounded in the theory of evolution by natural selection. Brandon, for example, describes his view as follows:

> [My view] is deeply committed to a certain view of evolutionary theory and the evolutionary process. If this view turns out to be mistaken then teleological explanations as I characterize them would have to be rejected. This is in contrast to what seems to be the current trend in the philosophical literature. This trend is to give accounts of *biological* explanations not tied to any specific biological theory. [Here he cites Wright 1976.] I can't see the interest in an account indifferent between divine creation and Darwinian evolution. My account does not aim at that level of abstraction. (Brandon 1981, reprinted in Brandon 1996, 30)

By eschewing an account indifferent between divine creation and Darwinian evolution, Brandon rejects the broad concept of selection to which Wright appeals and focuses instead upon the concept of selection as it occurs in evolutionary theory. This is not to reject the project of analyzing the concept of functions; it is merely to insist that we analyze the application of the concept in a relatively narrow range of contexts, namely, in the context of theorizing about the evolutionary history of life on earth.

22. See Chapter 2 for a causal rather than explanatory formulation of the theory.

23. Papineau (1993), for example, holds that selection occurs in the process of mental ontogenesis: ". . . selection-based teleology can also be the basis of individual learning" (Papineau 1993, 59). This line of thought goes back at least as far as Dennett (1974). The more specific suggestion is that even the functional status of artifacts is somehow constituted by the process of selection; see, for example, Griffiths (1993).

And restricting our analyses to the concept as employed among evolutionary biologists may give the appearance of naturalizing functions.[24]

Like Brandon, Neander (1991) restricts the scope of the theory of selected functions to the use of the concept "function" among modern biologists. Against Millikan (1989), Neander insists that a theory of functions requires an analysis of the concept. She rejects Millikan's restrictive characterization of conceptual analysis as the search for necessary and sufficient conditions for the application of the concept. Instead she allows that it may be fruitful to describe "those criteria of application that people standardly apply, most of the time, in the most standard contexts" (Neander 1991, 171). Such criteria often are vague, context sensitive, and variable across individual language users, to be sure, but this should make us cautious without being dismissive. More importantly, we can describe the criteria of application that experts within the life sciences employ in the course of theorizing. This, according to Neander, is likely to bear fruit, for when the concept under investigation is a term of art in a well-developed scientific theory, worries about variability and vagueness are greatly diminished:

> when the relevant linguistic community consists of specialists, and the term under analysis is one of their specialist terms, and is also abstract (nonperceptual) and embedded in well-articulated theory, the severity of each of these factors [vagueness, context-sensitivity, etc.] will be greatly reduced. Since this is the case for contemporary biologists and their notion of a "proper function," we have more reason to expect success in this case than in most. (1991, 171)

Since "function" is a term of art, referring to a concept embedded in a well-developed theory, Millikan's prohibition against conceptual analysis is unnecessarily restrictive. The aim of the theory of selected functions, on Neander's view, is to specify the conditions under which the concept is properly applied among modern biologists. There is no assumption here that biologists have an explicit account of functions woven into their biological theories. It is enough, on Neander's view, that "biologists implicitly understand '[selected] function' to refer to the effects for which

24. Brandon focuses on "What for?" questions as raised and as answered by evolutionary biologists. This amounts to explicating the concept of functions as employed by evolutionary biologists. See chapter 7, section IV, for further discussion.

traits were selected by natural selection" (1991, 176). It is this implicit understanding that the theory of selected functions is designed to describe.

The strategy, then, is to capture and depict conditions that lead contemporary biologists to attribute functions to natural traits. After all, biologists do employ the concept of functions in theoretical contexts. It is sometimes claimed that talk of functions in biology is ineliminable, that biologists are somehow forced or obligated by the phenomena to employ functional notions in order to express their claims. More importantly, biologists, in the process of learning their trade, hone their instincts and their intuitions, clarify their concepts, and distill their theories while studying the very systems we take to be functional. It is in the course of studying living things and the processes and mechanisms that make up living things that biologists sharpen their concepts and intuitions. It thus is reasonable to expect that the theoretical use of the concept "function" within biology is the most informed—the closest to the true nature of functions—to which we have access.

This line of reasoning, I take it, is a kind of Quinean argument for the methodology employed by advocates of selected functions. Advocates of selected functions make no explicit appeal to Quine, but the basic claim is that our ontological commitments should be guided and constrained by the postulations of our best science—in this case, contemporary evolutionary theory. Since contemporary biologists apparently postulate functional properties, it is reasonable to explicate the conditions that lead to such postulations in order to understand the nature of such properties. Brandon, Neander, and others reject Wright's method of inquiry on the grounds that it is untethered from the methods of any well-developed scientific inquiry. Analyzing Wright's generic concept of selection may reveal important features of that generic concept, but there is no particularly good reason for thinking that everyday uses of this concept orient us toward the truth about biological functions. There is, by contrast, reason for hope when we restrict our analysis to uses of the concept among those most intimately acquainted with the workings of living things.

This progress from Wright's etiological functions to the theory of selected functions illustrates the methods of the latter. The aim is to specify conditions under which contemporary biologists apply the concept of functions. The scope of the analysis is narrower than Wright's, to be sure, but the aim of explicating a specific concept remains the same. Having employed this strategy, advocates of selected functions announce their findings: Biologists, they claim, tend to apply the concept when they believe (implicitly or explicitly) that the specified trait performs the relevant task as a consequence of being selected for that task. Biological functions, they conclude, are properties that have been selected for. They further claim that, in the course of theorizing about living systems, biologists apply the related concept of malfunctions to tokens that fail to perform their typical functional task. And this leads us to the additional conclusion that selected functions are properties that persist in the face of physical incapacitation. These, then, are the methods of inquiry employed by the theory of selected functions.

Several doubts arise immediately; I will mention just two. The first is that these methods misappropriate Quine's restrictions regarding ontological commitment. Quine maintains that we should countenance in our ontology only those entities postulated in our best-developed scientific theories. The obvious cases are those in which unobservable entities are postulated in the context of a theory that otherwise successfully explains and predicts a wide range of observable phenomena. Quarks and bosons are postulated in order to explain certain physical phenomena and, to the extent the larger theory is compelling, we ought to believe (or at least accept[25]) that quarks and bosons exist. But it is doubtful that talk of functions in biology is analogous. Biologists from several subdisciplines employ functional terms, to be sure, but it is hard to believe that they posit unobservable functions in order to explain observable phenomena. It is hard to believe that certain observable biological phenomena would be rendered unintelligible were we to ban the attribution of selected functions from all biological theories. There may be some claims we could no longer justify—for example, that a damaged token is failing to do what it is "supposed to" do—but that does not show that the condition of that token would suddenly become inexplicable. At most it shows that we would have to reconceptualize the functional status of damaged

25. Where accepting the postulations of physicists is allegedly distinct from believing in the entities postulated. See van Fraassen (1980).

tokens. I suggest one such reconceptualization in chapter 6. Moreover, and as I argued in chapter 3, we can explain the workings of systems and their components by appeal to systemic functions, including the workings of populations affected by selection. This means that the alleged Quinean rationale may lead us to systemic functions but does not force us to accept the theory of selected functions. At any rate, the burden rests on advocates of selected functions to defend the applicability of the alleged Quinean strategy, for otherwise they have no grounds on which to stake the methods by which they approach functions.[26]

The second problem is more telling. The above methods fail to address a central concern of any theory of functions. The concern is to show how biological functions are consistent with the methods and postulations of the natural sciences generally. As emphasized at the outset (chapter 1), the physical and chemical sciences endeavor to account for what there is and for what is possible. The aim is to explain why things are as they are. The biological sciences aim to explain more. In addition to explaining why things are as they are, they also aspire to explain why some things are supposed to be one way rather than another. The aim is to explain the norms of nature and to help us understand the nature of functions in a way that fits well the methods and postulations of all the natural sciences. And, also emphasized at the outset, it is naive to think that biologists are especially qualified for this job. Biologists may be attuned to living systems; they may have sharpened their intuitions and instincts in the course of studying functional systems. But that is hardly sufficient for the task of reconciling the attribution of functions with the methods and postulations of the natural sciences generally. Indeed, the task before us exists precisely *because* the concept, so fruitful in biology, fits so poorly a wider naturalistic orientation toward inquiry. A mere description of the problematic concept or the conditions under which this concept is applied is bound to miss the mark. The allegedly Quinean approach may enable us to better understand the conditions under which the problematic concept is employed, but it is difficult to see how it enables us to answer the

26. Just to be clear, none of the advocates of selected functions of which I am aware has appealed explicitly to this alleged Quinean argument. I offer it as a charitable take on the methods of the theory.

central question concerning the naturalistic credentials of such norms. So, the methods of the theory of selected functions fail to address a central question facing any theory of functions.

IV Confirmation and the Ontology of Selected Functions

But suppose for the sake of argument that the general methods of the theory are unobjectionable. Nevertheless, a significant problem remains. The problem is that the ontology of the theory is unacceptable from a naturalistic perspective. To show this, I will not insist that we commit ourselves in advance to a given ontology. The strategy, rather, is to show that there are important methodological reasons for rejecting the ontology of selected functions. My objections rest upon a few rather bland assumptions concerning the best methods for inquiring about the natural world. To begin, I describe what I regard as the two most plausible options concerning the ontology of selected functions. I then argue that, on either construal, the naturalistic credentials of selected functions are wanting.

The question is, What are the ontological commitments of the theory of selected functions? What kind of property are we attributing to a token trait when we attribute a selected function? More generally, where in our naturalistic landscape do selected functions reside? A minimalist about selected functions may hold that a token trait possesses selected function F exactly when it is descended from ancestral tokens that were selectively successful by virtue of performing task F. On this view, possession of a selected function is equivalent to possession of a certain kind of history—a history of selective success. Nothing about the current intrinsic physical properties of the token is relevant; the right sort of historical relation is everything. Millikan is a clear exponent of this view: ". . . being preceded by the right kind of history is *sufficient* to set the norms that determine purposiveness; the dispositions themselves are not necessary to purposiveness" (Millikan 1989, 299). A more robust view, by contrast, may hold that a token possesses selected function F exactly when it is descended from ancestral tokens that were selectively successful by virtue of doing F *and* when it currently possesses the physical properties required to perform task F. On this view, the right sort of historical relation is necessary but not sufficient; current capacity is additionally required.

The first option offers the more charitable interpretation of the theory of selected functions.[27] The second option requires possession of the requisite physical capacity and this has the unhappy effect of ruling out the possibility of malfunctions. Token traits that are sufficiently damaged or diseased, and thus devoid of the requisite physical capacity, would no longer qualify as functional. They would have to be classified as nonfunctional. And nonfunctional tokens no longer belong to a functional category, in which case they cannot qualify as malfunctional. We thus should adopt the first option—minimalism—in assessing the ontology of the theory.

Minimalism may appear persuasive, at least at first glance. The minimalist holds that possession of a selected function is equivalent to possession of the right sort of selective history. And as we saw in chapter 2, it seems intuitive that the sorting capacity of selection produces lineages of functional properties. The idea is that past selective success somehow shapes subsequent generations, somehow determines what descendent tokens are for. Current tokens are "for" the performance of just those tasks that contributed, by way of selective success, to the persistence of the trait type. This is to say that historical relations of the right sort—those concerned with selective success—manage somehow to fix the office or role imposed upon subsequent generations. Possession of a selected function, on this view, involves nothing more than possession of this sort of historical relation. That, at any rate, is the intuition that gives the theory of selected functions its prima facie appeal. And a minimalist ontology appears sufficient to support the intuition.

Upon reflection, however, minimalism is untenable. For it is incumbent upon advocates to tell us how selected functions are constituted out of

27. While perhaps a charitable take on the ontology of selected functions, this interpretation nevertheless is incorrect. In chapter 7, I argue that selected functions require both (a) and (b)—both the right sort of historical relation and the requisite physical capacity. For present purposes, my strategy is to allow advocates of selected functions the least-burdensome ontology consistent with the thesis that functions are norms that attach to token traits. I intend to show that, even with a forgiving ontology, the commitments of the theory are unacceptable.

the mechanisms involved in natural selection. A naturalistic account of functions demands at least that much. Yet it is doubtful that this burden can be met. Consider first the leading intuition behind minimalism, namely, that natural selection has the power to shape, determine, and impose functional offices and roles. If we keep our attention fixed upon nonengineered, natural traits—if we avoid conflating claims about artifactual functions with claims about natural functions—this intuition is puzzling, even mysterious. As naturalists we must ask, *What in the process of natural selection* makes it true that descendent tokens are "for" the performance of a given task? What in the process of past selective success determines, shapes, or imposes such functional roles? What natural features of the causal-mechanical processes that constitute a selective history have the power to determine that descendent tokens are for the performance of some task? Advocates of selected functions have not addressed these questions, but they should. It is *their* claim that the theory of selected functions naturalizes our concept of functions. The mere assertion that selected functions are equivalent to selective success is perhaps a plausible opening line, but it cannot be the whole story. As naturalists, we require more than the bare assertion of the view. We require an *account* of the natural mechanisms or the natural causes that give rise to the functional offices and roles. We expect, for example, an account of the natural selective process that manages to impose upon descendent hearts the role of pumping blood. We expect to be told what in the natural world accomplishes this. But we have yet to be told. No advocate of selected functions has given a plausible explanation of how, or by what means, the process of selection gives rise to such roles. This part of the theory is asserted but never defended.[28]

And upon reflection it is hard to imagine what advocates *could* offer by way of defense. Natural selection occurs when variation among types of traits produces differences in reproductive output. It is hard to see how differential reproduction resulting from variation among traits makes it the case that descendent tokens are "for" the performance of some task.

28. In consequence, attempts to naturalize other sorts of phenomena by appeal to the alleged normative status of selected functions and malfunctions are, to put it mildly, implausible. Dretske (1995) on consciousness is typical.

It is hard to see, from a naturalistic perspective, what in the process of selection could assign such functional roles. What is the naturalistic story? What are the relevant natural mechanisms and how do they work? Selective success may prompt us to *expect* descendent tokens to carry on in the manner of their successful ancestors, but that is a fact (if it is a fact) about the effects of selection on our psychology.[29] Selected functions, by contrast, are supposed to be properties that result from the success of ancestral tokens and attach to descendent tokens; the effects of selective success on our psychology are irrelevant to the theory of selected functions. At the very least, the burden rests on advocates of selected functions to fill this gap in their attempted naturalization.

Advocates of selected functions, beginning with Brandon, Neander, and Millikan, inherit the burdens of naturalism. It is not enough to assert that the notion of functions among life scientists refers somehow to natural selection. Such an assertion may serve as an analysis of the concept as employed by biologists, but conceptual analysis even in this restricted form fails to point to the relevant natural processes or mechanisms responsible for the emergence of functional properties. The alleged naturalization has not yet been given.

But the problem for selected functions is worse still. Minimalism is an internally conflicted view. The conflict is perhaps clearest in Millikan (1989). She claims that possession of a selected function involves nothing more than possession of a selectively successful history and, at the same time, that traits with selected functions can, if sufficiently incapacitated, malfunction. But she cannot have it both ways. If minimalism is the correct ontology of selected functions, then selected malfunctions are impossible; and if selected malfunctions are possible, minimalism is false. To see this, consider the claim that some token trait is malfunctional. The claim is that a damaged or diseased token, devoid of the requisite physical capacity, nevertheless is supposed to perform its functional task. This is a remarkable property—being such that the token is supposed to do something that it physically cannot do—and we must wonder what in the natural world could give rise to this property. If we assume that selected

29. I speculate on the importance of such expectations in section III of chapter 6.

functions are equivalent to past selective success—if we adopt minimalism—then advocates must tell us what mechanisms in the process of natural selection make it the case that a damaged heart is supposed to do something that it is physically unable to do. As we saw in chapter 2, advocates hold that the emergence of selected functions involves the emergence of a functional office or role, including a norm of performance that applies to tokens of the functional type, a norm that remains even when the requisite capacity is lost. So what in the process of natural selection is responsible for the emergence of such norms? What causal-mechanical properties of our history have the power to produce norms that attach to descendent tokens and remain attached even when tokens do not possess the physical capacities required to fulfill the norms?

I doubt that advocates of selected functions can give suitably naturalistic answers to these questions. There are no mechanisms in the processes that comprise natural selection endowed with the power to produce such norms. Or, more modestly, advocates of selected functions have yet to tell us what the relevant mechanisms are and how they work. So, it seems to me that advocates either must admit that they have no account of the natural processes that give rise to such norms, in which case they have failed to naturalize the attribution of malfunctions, or, in order to preserve the attribution of malfunctions, they must adopt an ontology that is difficult to characterize in naturalistic terms. Now, opting for the first horn is plainly unattractive. It enables advocates of selected functions to retain the minimalist ontology, but they cannot account for the possibility of malfunctions. And since accounting for malfunctions is something most advocates of selected functions aspire to achieve, it is doubtful that the minimalist ontology can survive. This drives us toward the second horn.

Opting for the second horn involves rejecting the minimalist ontology for a much fatter account of selected functions. On this view, selected functions are equivalent not to past selective success, but rather to the roles or offices that selective success imposes. They are the *products* of past selective success and, as such, they are not equivalent to the physical *processes* that produced them. Moreover, insofar as we are warranted in attributing malfunctions, selected functions are not equivalent to any physical properties of descendent token traits, since the function persists even in the face of physical incapacitation. Characterized in this way, selected functions are surprisingly rarefied properties. Indeed, they are abstract and noncausal properties. They must be abstract because they are not equivalent to any historical or contemporary physical properties of the organism. And they are noncausal on the assumption that nonphysical properties are causally impotent, and also on the assumption that a selected function persists even when the token is causally unable to perform the associated task.

The problems with this second option, from a naturalistic perspective, are all too obvious. It is prima facie difficult to square the postulation of abstract, noncausal properties with the methods and postulations of the natural sciences generally. Nevertheless, the concern to account for the possibility of malfunctions drives us away from minimalism toward this sort of ontology. And perhaps advocates of selected functions have room to maneuver. They may point out what several philosophers have already noted, namely, that scientific inquiry seems to require the existence of other sorts of abstract properties and entities, including numbers or sets. "Why, then," they might ask, "worry about the postulation of one more kind of abstract property? If positing selected functions conduces to inquiry—as does the positing of numbers or sets—why let our ontological conscience get in the way? Why not carry on and see how far we can go?"

At least three considerations, based on naturalistic methods of inquiry, tell against this defense of selected functions. (1) Biological theory does not require the positing of abstract, noncausal properties, in which case the alleged analogy to numbers or sets does not hold. Science may require mathematics, but biology does not require the sorts of norms postulated by the theory of selected functions. Indeed, unless biologists have conspired to conceal from the public some extraordinary phenomena, it should be clear that abstract, noncausal norms of performance are extraneous on explanatory and predictive grounds. Darwin on the evolution of the eye (see the second epigraph to this book) illustrates the fact that evolutionary biology has no need of such norms. Darwin offers a compelling explanation of the evolution of a complex and adaptive trait, and he does so without attributing properties that are abstract and noncausal. To add to Darwin's explanation the claim that descendent eyes now possess abstract, noncausal norms of performance is to go beyond the scope of the evolutionary explanation. It is to engage in a kind of theorizing that Darwin evidently saw no need to pursue. Biological theory does not somehow warrant the positing of such norms.

Of course, several biologists have attempted to confer some degree of respectability upon talk of natural norms in biological theory. Ayala (1970), for example, insists that biological explanations, unlike explanations in nonbiological sciences, require appeal to selected functions. Such explanations, he claims, insofar as they appeal to natural selection, commit us to the existence of teleological properties:

> Natural selection can be said to be a teleological process in two ways. Firstly, natural selection is a mechanistic end-directed process which results in increased reproductive efficiency. Reproductive fitness can, then, be said to be the end result or goal of natural selection. Secondly, natural selection is teleological in the sense that it produces and maintains end-directed organs and processes, when the function or end-state served by the organ or process contributes to the reproductive fitness of the organisms. (Ayala 1970, 10)

I agree that reproductive fitness is a systemic capacity of organisms we wish to explain. And I agree that, in the context of this stipulation, we can analyze organisms into functional components. But this, on my view, is to say merely that evolutionary biologists can analyze populations and organisms with respect to fitness in terms of systemic functions. Beyond the attribution of such systemic functions, however, there are no grounds for the postulation of "ends" or "end-directed processes." So if Ayala is claiming that reproductive fitness is a genuine telos and not just a systemic capacity in which we have an explanatory interest, we should be skeptical. If he is claiming that the effects of hearts and eyes are in fact end-directed or that hearts and eyes possess abstract, noncausal norms of evaluation, we should resist. We should resist because evolutionary explanations simply do not require such goals, ends, or abstract, noncausal norms. Evolutionary explanations cast in terms of such norms may be convenient shorthand, but nothing in the substance of evolutionary theory requires that explanations be cast in such terms.

(2) To the extent that talk of functions is integral to biological theory, systemic functions suffice; we thus should not embrace an ontology of abstract, noncausal properties since a more suitably naturalistic alternative is readily available. To see this, notice first that abstract, noncausal

norms are at odds with the postulations and the methods of other natural sciences. The methods and postulations of other natural sciences are relevant on genuinely Quinean grounds—not the alleged Quinean argument given above—for when Quine said that our best sciences are our best guide to what there is, he surely meant our best sciences *taken as a whole.* He did not intend that we fix our attention upon the use of a single notion (or a single cluster of notions) within a single domain. Our best confirmed beliefs form a web within which the implications of any one belief reverberate throughout the whole or at least throughout a nontrivial portion; the postulations and methods of our natural sciences generally are woven closely together in this web. In assessing the theory of selected functions, we thus should take note that physicists and chemists refrain from attributing properties that are nonphysical and noncausal. They do so, of course, because it is difficult to test for the existence of nonphysical postulations and because it is hard to see what theoretical role noncausal properties can play.

There is, of course, the point mentioned above about numbers and sets. Science seems to have need of such things, so perhaps we must admit their existence. Perhaps.[30] But that line of reasoning does not apply in the case of selected functions. This is because we have a highly plausible alternative theory of functions. This alternative is preferable to the theory of selected functions on grounds independent of ontological commitments—recall the arguments from chapter 3—but with respect to ontology, the theory of systemic functions is clearly superior. The theory of systemic functions posits the existence of systemic capacities only when they are grounded in physical mechanisms within the relevant system. The ontology of systemic functions is thoroughly naturalistic and eschews the postulation of abstract or noncausal properties in nature (more on this in chapter 6). An analogous point applies to the postulations of physicists and chemists. Natural scientists surely are right to resist the postula-

30. Field (1980) and Papineau (1993) challenge the claim that numbers must be conceptualized as abstract, noncausal entities. Numbers, they suggest, are useful fictions. If such fictionalism is plausible, then the importance of numbers to scientific inquiry can be acknowledged even while denying that numbers or sets are part of our ontology. In that case, if other things are equal, naturalists ought to be fictionalists. See Papineau (1993), chapter 6, for discussion.

tion of abstract, noncausal properties whenever there are other properties available—physical and causal in nature—that explain the relevant phenomena. So, we can allow that a total prohibition against abstract or noncausal properties is too blunt a theoretical instrument, but we can also insist that it is better to prefer the theory that avoids such properties so long as other things are more or less equal. And that is how it is with respect to selected functions: Given the availability and power of systemic functions, selected functions ought to be avoided.

(3) There are, moreover, historical precedents that tell against the postulation of such norms, even when postulated by professional biologists. Consider the many deists among Darwin's contemporaries. Among these were some of the best scientific minds of nineteenth-century England: William Whewell, John Herschel, Charles Lyell, Richard Owen, and others. These scientists in one way or another appealed to the concept of divine creation or divine design in their theoretical works. Premises of their theoretical arguments appeal to God's plan, implemented in the creation of the universe and embodied in the laws of nature or, to use Herschel's (1831) notion, embodied in the *vera causae*.[31] These thinkers continued to postulate the existence of divine design long after the publication of Darwin's theory. Nor were these theorists in the minority. They were in the majority within the scientific community and occupied the most powerful scientific posts.[32] It thus is plausible that, had we surveyed

31. Ruse (1979) describes Herschel's notion this way: "On the one hand we have mere *empirical* laws—laws that connect things and point to regularities without really showing why things occur as they do. Paradigms of this case are Kepler's laws, which prove the regularities of the planets without really explaining why such regularities exist. But the aim of the scientist must be to explain the reason for empirical regularities, and this involves considering higher laws—laws that refer to *causes*. Regretfully, Herschel was anything but precise about what he meant by 'cause,' but essentially he seems to have had in mind the idea of one phenomenon, the cause, in some way leading to or 'making' occur another phenomenon, the effect. . . . For Herschel the highest form of cause was force—indeed, he suspected that all cause was in some way reducible to force. . . . Moreover, Herschel suspected that all force was will-force . . . if not man's, then presumably God's" (Ruse 1979, 57).

32. See Hull (1973) and Gillespie (1979). I do not claim that, at the time of writing the *Origin,* Darwin had rejected all religious belief. I claim only that he deals with deism in the *Origin* as I describe above. Gillespie's discussion as a

English naturalists during the period from 1830 to the turn of the century, we would have discovered that most of them *explicitly* postulated the existence of some form of intelligent design in the natural world. We would have discovered that such postulations, according to these scientists, were required in order to account for certain features of the phenomena being explained. We thus would have been driven, in the face of the allegedly Quinean rationale for selected functions, or at least in the face of the point about numbers and sets, to adopt some form of deism.

But of course we now accept that Darwin was right to challenge their postulations. We now accept that, on the basis of the evidence available to him at the time, Darwin was right to reject the appeal to divine design. And at least three of the considerations raised by Darwin against his deistic peers also tell against the theory of selected functions. (a) Advocates of selected functions do not appeal to divine design, of course, but they do appeal to some sort of "natural design," part of which includes historically derived norms of performance. Now, Darwin repeatedly argued that all of the relevant biological phenomena, including the complexity and adaptedness of the mammalian eye, can be explained in terms of (mainly) selection and some as-yet-undiscovered mechanisms of inheritance. The deistic hypothesis added nothing and thus was explanatorily extraneous. Likewise, we should agree that *selected functions (abstract, noncausal norms) are explanatorily extraneous.* Our evolutionary explanations of complexity and adaptedness have no need of such norms. Nor do our explanations of systemic functions. As the example of the eye makes clear, natural selective explanations require no such norms; Darwin explains the evolution of the eye without postulating any norms of performance. And if the theory of systemic functions, as developed throughout this book, is compelling, then we can generate a full range of function attributions, including those that emerge in systems subject to selection, from within that theory. We can explain why a type of trait has persisted by appeal to systemic functions alone. The theory of selected functions, construed as an autonomous theory, adds nothing; we should reject it as surely as Darwin rejected the doctrine of creation on grounds that it was

---

whole confirms this claim. Gillespie's discussion in chapter 8 takes up the question of Darwin's religious beliefs prior to and following publication of the *Origin.*

extraneous. (b) In resisting the postulation of divine design, Darwin also complained that appeals to such design were beyond the reach of scientific confirmation. Appeals to divine design invariably failed to specify secondary causes that can be studied by human beings. At crucial points in their explanations, deists appealed to the wisdom and efficacy of divine design but failed to specify the earthly mechanisms responsible for implementing this design. This had the effect of insulating claims about design from evidential test. The appeal to natural design is no better in this regard. Selected functions are abstract and noncausal in character; if not, it is hard to see how selected malfunctions are possible. Yet it is perhaps harder to see what sort of scientific inquiry could confirm or disconfirm the existence of such norms. Darwin's complaint about the lack of secondary mechanisms is every bit as pressing in the theory of selected functions. Thus, we should reject the theory on the grounds that *selected functions (being abstract and noncausal) are beyond empirical investigation.* At the very least, the burden rests on advocates of selected functions to show that the postulations of their theory are subject to reasonable standards of test. (c) Finally, if both (a) and (b) are compelling, we should conclude that *selected functions, being theoretically extraneous and beyond reasonable test, ought to be discarded from a naturalistic perspective.* This is how Darwin dealt with deism. Deism is theoretically irrelevant given the alternative suggested by Darwin. It is also beyond scientific test. Perhaps it redescribes the phenomena in terms that are familiar, or terms that, because of their institutional and historical roles, provide some type of comfort. But within the context of understanding the world we inhabit in naturalistic terms, deism is not a hypothesis with anything to recommend it. Parallel considerations apply to the theory of selected functions; it is a theory to which naturalists should turn their backs.[33]

33. It is perhaps worth mentioning a recent work in which the theological undertones of selected functions are given full voice. Plantinga (1993) assumes a rather general notion of proper function and argues that attempts to naturalize that notion cannot succeed. On the basis of that argument, he concludes further that proper functions should be understood in *supernatural* terms instead of naturalistic ones. Now, I agree (though for different reasons) with Plantinga that the theory of selected functions has naturalistic credentials that are dubious at best. So

I conclude that the theory of selected functions, taken as a whole, is insufficiently naturalistic. Neither the methods nor the postulations of the theory rise to the standards of a naturalistic approach to inquiry; neither fit comfortably the methods and postulations of the natural sciences generally. There is irony in this. Advocates of selected functions reject the generic analyses of Wright (1973, 1976) and opt instead for analyses of expert use of the concept of functions, giving the appearance of a cozy relationship between selected functions and evolutionary theory. But advocates also want more than evolutionary theory can provide. They want a theory of functions that is normative, a theory that accounts for the intuition that some functions are "proper" or "teleological." They want a theory of functions that underwrites the possibility of malfunctions. The only way to get what they want is to hold that functions are offices or roles with built-in norms of evaluation, in which case they are equivalent to neither the selective history of the functional trait nor the physical capacities of the organism. And once we admit an ontology of abstract, noncausal properties, the theory of selected functions falls into conflict

---

something has to give. But what should we be most willing to relinquish—our commitment to naturalism or to the existence of "proper" functions? Given the great advances garnered by the methods of our natural sciences, and given the mostly metaphysical history of our concept of functions and related notions, there is no real contest here. Adverting to supernaturalism in order to retain the notion of proper functions is troubling in at least two ways. First, it lacks all sense of proportion. It commits us to a cosmic theory in order to accommodate an intuition concerning highly local phenomena—an intuition, moreover, that we can explain away quite neatly. Second, adverting to supernaturalism fails to take seriously the evident power of scientific methodology—and it is methodology, not ontological prejudice, that motivates the sort of naturalism endorsed here. Our metaphysical postulations should come *after* exhaustive inquiry; they certainly should not constrain inquiry. Naturalism, that is, is first and foremost about the most promising methods for acquiring truth about reality; our ontological commitments—including what Plantinga calls "metaphysical" naturalism—should be constrained by those methods. And, as I have been arguing, the methods of natural science have no need of "proper" functions; systemic functions are more than enough. (See chapter 6 for further elaboration of this point.) At any rate, Plantinga's view makes clear the theological baggage that our concept of functions tends to drag in its wake. It is no surprise that Darwin's more-general criticisms of nineteenth-century creationism apply to contemporary theories of allegedly "proper" functions. Advocates of selected functions should hesitate to complain that my appeals to Darwin are out of place.

with the methods and postulations of our best sciences—*including* the methods and postulations of Darwin's theory of evolution! The theory of selected functions is defeated, in part, by its failure to mesh with what is supposed to be its closest ally among the natural sciences.

Brandon (1981) asserts that if evolutionary theory is false, then so too is the theory of selected functions. This may give the impression that selected functions are integral to evolutionary theory and thus naturalistic in character. But this is a misleading impression. Selected functions are not part of evolutionary theory; they are an add-on. The truth of evolutionary theory is hardly sufficient for the postulations of abstract, noncausal norms of performance. We thus can reject the theory of selected functions without broaching the status of evolutionary theory. To establish a close link between selected functions and evolutionary theory requires an argument to the effect that the truth of evolutionary theory makes plausible the reality of selected functions. My view, to the contrary, is that the truth of evolutionary theory, combined with the truth of theories from other areas of inquiry, makes the reality of selected functions highly implausible. The sorts of considerations that drove Darwin to reject divine design should drive us to reject "natural design," including selected functions.

All of this should bring the thesis of chapter 3 into greater focus. When I say that the theory of systemic functions can account for all the function attributions warranted within the theory of selected functions, I certainly do not mean that systemic functions that arise in the course of selection are abstract and noncausal norms. On my view, natural traits cannot malfunction; they can disappoint our expectations, but they cannot malfunction (more on this in chapter 6). For, systemic functions are nothing more than systemic capacities individuated within a given analysis of a given system. This does not diminish the theory of systemic functions. Indeed, the theory can explain how a population changes or remains the same; it thus can explain, in the context of the workings of a population, why a trait has persisted or proliferated. But it in no way tries to explain what a trait is "for." We should reject the claim—a claim central to the theory of selected functions—that an explanation of why a trait exists is equivalent to an account of what a trait is for. There are good accounts of why a trait has persisted, but there are no good accounts of what natural traits are for. This is also why I reject the views of Griffiths (1993) and others, which attempt to preserve the attribution of malfunctions within the theory of systemic functions.

If the arguments of this and the previous sections are sound, we have reason to reject the intuition that some natural traits are "properly" functional or genuinely "teleological." We should hold, rather, that some systemic functions arise in the course of selection and others do not, and those that do are on an ontological par with all other systemic functions. Selection does not transform mere systemic functions into abstract, noncausal norms of evaluation. This view of the matter, however, leaves open the question, Why are we inclined to attribute such norms of evaluation? Why are we inclined to see some traits as functional and others as merely useful? Or, less contentiously, why are we inclined to see some systemic functions as more central than others? To round out the discussion of this chapter, I will speculate briefly on the possible sources of our inclination.

V Complexity

The epigraphs to this book contrast the confidence of Hume's deist with Darwin's measured insistence that purely causal-mechanical processes can produce complex adaptations. Today we accept Darwin's view over deism; today any appeal to divine design in the course of biological inquiry is theoretically unmotivated. Nevertheless, the intuition to which deists appeal—the sense we have that living systems and their parts are indeed purposive—is one that Darwin never shook off entirely. In 1881, some twenty-two years after the publication of the *Origin,* Darwin had the following exchange with the Duke of Argyll, preserved by Emma Darwin and described here by Gillespie:

> When the Duke of Argyll told the aging scientist that it was impossible to look at the numerous purposeful contrivances in nature and not see that intelligence was their cause, Darwin "looked at [him] very hard and said, 'Well, that often comes over me with overwhelming force; but at other times,' and he shook his head vaguely, adding 'it seems to go away.'" Design was the nagging doubt that never left Darwin's mind. (Gillespie 1979, 88; quoted passage from Litchfield 1915)

Despite having shown that appeal to an intelligent mind is extraneous and beyond test, and hence contrary to a naturalistic orientation, Darwin was nonetheless struck from time to time by the apparent purposiveness of living systems.

It seems to me that many of us are similarly struck today. At least some advocates of selected functions struggle to explicate the sense we have that living things and their parts are, in some sense of the term, designed or purposive. They are so struck despite rejecting any appeal to an actual designer within the natural realm. In the end, advocates of selected functions struggle to account for the apparent purposiveness of living things by postulating properties that conflict with our naturalistic commitments. They appeal, in the end, to norms of performance that are abstract, noncausal, and persist in the face of physical incapacitation. The most obvious alternative to such norms is the minimalist ontology described above, but minimalism leaves no room for a naturalistic account of the possibility of malfunctions. In the interests of avoiding this dilemma, I wish to suggest a quite different explanation for our sense that the biological is purposive. I wish to consider the question, Why do some things elicit in us the intuition that they are "for" the sake of some end?

My suggestion—which is tentative, partial, abstract, and entirely speculative—is that, as hierarchical systems increase in internal complexity, the attribution of systemic functions becomes more compelling. This is a psychological hypothesis. Complexity in a hierarchical system strengthens our conviction that certain systemic components are endowed with systemic functions. Hierarchical systems are internally complex to the extent that relevant lower-level capacities, in producing specified higher-level capacities, are *causally interdependent*. The mammalian circulatory system, for example, is not only hierarchical and interactively cohesive, but also complex insofar as the capacity of one component is causally interdependent with those of several others. The heart pumps only when it is supplied sufficient oxygen; the heart is supplied sufficient oxygen only when the lungs work; but the lungs work only when the heart pumps. The efficacy of the heart is interdependent with that of the lungs and several other components. The circulatory system contrasts sharply in this regard with Haugeland's fiber-optic cable. If one or a few individual fibers within the cable cease to produce their effects, the other fibers can carry on as before. But if one or a few components within the circulatory system cease, the entire system shuts down; all the other components, being causally dependent, promptly cease to produce their effects. To the extent the components of a hierarchical system are causally interdependent in this way, we are inclined, I believe, to see them as having the "job" or the "role" of producing such effects. We are inclined, that is, to see these components as "for" the production of certain effects. After all, we find it irresistible to conceptualize the heart in terms of a job or role, but challenging to conceptualize individual fibers of the optic cable in such terms. The reason, I wish to suggest, is that the heart is causally and systemically interdependent in ways the optic fibers are not.

But why are we inclined in this way? Systems that are complex and hierarchical have two notable features. The first is that they tend to be stable over time. Simon (1973) contends that hierarchical organization is pervasive in nature because interactions at the lowest levels of organization are governed by the strongest bonding relations and thus tend to be the most cohesive. Natural forces governing atomic structures, for example, are stronger than those governing the structure of large macromolecules. This accounts for the hierarchical structure of various chemical elements:

> protons and neutrons of the atomic nucleus interact strongly through the pion fields, which dispose of energies of some 140 million electron volts each. The covalent bonds that hold molecules together, on the other hand, involve energies only on the order of 5 electron volts. And the bonds that account for the tertiary structure of large macromolecules, hence for their biological activity, involve energies another order of magnitude smaller—around one-half of an electron volt. It is precisely this sharp gradation in bond strengths at successive levels that causes the system to appear hierarchic and to behave so. (Simon 1973, 9; quoted in Bechtel and Richardson 1993, 29)

Significantly, if natural hierarchical systems tend to be highly cohesive at their lowest level of organization, tokens of the relevant systems will tend to persist over time. Their persisting over time will have the further effect that the higher-level capacities of such systems will be relatively stable as well. And this is related to the second notable feature of complex, hierarchical systems, namely, that they tend to perpetuate themselves in at least two ways. Token systems tend to produce internal states that persist or recur; they also tend to reproduce their kind. Organisms perpetuate themselves in both ways. Organisms are systems that tend to persist because of the causally interdependent, typically stable effects of their own components.

Many hierarchical systems—perhaps the majority—tend to be stable and self-perpetuating. There is no conceptual necessity in this, but it seems so all the same. My suggestion, then, is that certain of our psychological capacities and limitations incline us to conceptualize the capacities of stable, self-perpetuating systems as especially functional. In particular, those capacities integral to the system's stability or self-perpetuation strike us as especially salient systemic functions. They strike us as being for the sake of various tasks that contribute to or result from the stability of the system. After all, experience teaches that a defect in even a single component can impede the workings of complex, hierarchical systems. The study of medicine relies on this fact. We come to know, for example, that the stability of the organism—the working relations among its various systems—depends upon the efficacy of the heart. We learn that, if the heart ceases to pump or pump in the right way, the circulatory system is likely to falter and fail, making life difficult for the organism. As we acquire this sort of counterfactually-based knowledge about the interdependent parts of a given system, we find it tempting to conceptualize these parts as "for" the production of those effects integral to the stability of the organism. Precisely this type of knowledge inclines us to conceptualize the heart as endowed with quite specific functional roles. And it is hardly surprising that, among human beings and other animals of concern to us, survival and reproduction are stabilizing capacities in which we take a serious interest.

Exactly why we are struck in this way is not for me to say—that is up to the psychologists. My speculation is simply that we are so struck. Cast quite generally, the speculation is that there is a mesh between complex hierarchical systems and the cognitive capacities and limitations we bring to our study of such systems, a mesh that explains why the theory of systemic functions works. The theory of systemic functions is tailored to the kinds of inquirers we are and the kinds of phenomena we wish to investigate. The theory thus provides an approach to understanding functions from within the context of scientific inquiry generally.

The speculation that complexity in hierarchical systems causes us to see systemic components as especially functional—as "for" the production of certain effects—has significant consequences. Perhaps the most important is that it diminishes the intuition driving the promiscuity objection. That objection assumes that some natural objects are "properly" functional. As I argued in the previous section, the ontological commitments of this view are indefensible. Nevertheless, there is something to the intuition that some traits appear to us more functional than others. This is where the above speculation may assist us. Some systemic components seem to us especially functional because they operate within especially complex hierarchical systems that are stable and tend to perpetuate themselves. Components within highly stable systems strike us as being "for the sake of" the system's own stability or self-perpetuation, or "for the sake of" capacities that depend upon such stability. This fact about our reaction to complex systems is, I suggest, the source of our intuition that some traits are more functional than others. The intuition that some traits are "properly" functional is illusory, but the intuition that some are especially important within the system's internal economy is both real and true.[34]

A second consequence of the above speculation concerns an important difference between my view and the view of Nagel (1977). Our views overlap, but one difference is especially salient. Nagel casts his account of functions within a more-general account of goal-directedness, and he explicates goal-directedness in terms of certain systemic properties, particularly plasticity and persistence. A system is goal-directed if it has the capacity to persist in a stable state by means of various strategies, each of which applies under distinct environmental perturbations. The thermoregulatory system, as mentioned in chapter 2, illustrates Nagel's notion. It is goal-directed insofar as it maintains body temperature by means of various mechanisms. Some mechanisms—those that cause shivering—alter body temperature when the temperature of ambient air goes significantly below that of the body. Other mechanisms—those that cause perspiring—adjust body temperature when ambient air temperature goes above body temperature. The thermoregulatory system thus regularly maintains a rather narrow range of body temperature (persistence) by diverse means (plasticity). It is therefore appropriate, on Nagel's view, to attribute functions to the salient effects of the mechanisms involved.

34. The value of the above speculations resides less in the particular suggestions and more in the general Humean strategy of explaining away what may appear to some to be a powerful intuition by appeal to certain features of our psychology.

I agree with Nagel that functions are properly applied to systemic components. But I disagree that goal-directedness, as explicated by Nagel, is a necessary condition for the proper attribution of such functions. Haugeland's fiber-optic cable and other such devices convince me of this. I take it for granted that the cable is not goal-directed in Nagel's sense; it does not employ plastic means that result in the persistence of some state or some output. Nor is the cable complex in my sense of that term. This is why, in chapter 4, I develop the theory of systemic functions in terms of hierarchical organization, but not in terms of internal complexity. Hierarchy is necessary but complexity is not. Haugeland's cable can be analyzed into systemic structural capacities that give rise to its capacity to transmit images. This is a consequence of the fact that the cable is hierarchically organized, even if the levels of organization are few in number and simple in structure. There thus exist systemic functions in the absence of goal-directedness. So I suggest we think of goal-directedness as important but not necessary for understanding the nature of functions. It is important insofar as it accounts for our temptation to see some objects as more functional than others; but it is not among the conditions necessary for the attribution of a systemic function.

We can come at the same point from a different angle. Nagel claims that the function of any trait is to do F so long as its doing F contributes to the "welfare" of the larger system, where a system's welfare concerns contributions to systemic goals. The heart, for example, by pumping blood, contributes to the circulatory system's goal of delivering nutrients, which contributes to the organism's goal of maintaining a certain level of energy. The heart thus has the function of pumping blood relative to the welfare of the circulatory system and the organism's health. Functions, then, are relative to some systemic goal. Now, on my view, functions are indeed relative. The crucial difference, however, is that they need not be relative to any systemic *goal* explicated in terms of systemic welfare. They may, but they need not. On my view, functions are relative to some *capacity* of the system, but not all capacities are goals in Nagel's sense. Not all systemic capacities contribute to an organism's welfare; the reduced mobility caused by the narwhal's large tusk is a case in point (chapter 4). And not all systemic capacities are persistent or exercised by mechanisms that are plastic. I assume, for example, that the individual fibers in Haugelands's cable contribute to the transmission of some image in one and only one way. The efficacy of each fiber is not causally interdependent on that of the others. And nevertheless, relative to the larger cable, the systemic function of each fiber is to transmit a certain quantity of light. So it is false that function attributions to the individual fibers are goal-relative in Nagel's sense; they are merely capacity-relative. Of course, insofar as the capacities we study concern the genuine or apparent goals of a system, then capacity relativity and goal relativity overlap. But our inquiries into systemic functions are not so limited. Nor should they be.[35]

VI Conclusion

Advocates of selected functions would have us believe that the turn away from Wright's (1973, 1976) generic concept of selection toward the theory of evolution by natural selection somehow puts the theory of selected functions on secure naturalistic ground. But if the arguments of this chapter are sound, the naturalistic standing of selected functions is illusory. The ontology of the theory conflicts with the methods and postulations of other natural sciences, and attributions of selected functions are internally problematic and evidentially suspect. It is difficult to find any compelling signs of sound naturalistic credentials; it is difficult to agree with Neander that selected functions are embedded in any substantial way in contemporary biological theory. All of this, I conclude, provides additional reason for turning our backs on the theory of selected functions. By contrast, the theory of systemic functions is free from the defects that plague the theory of selected functions. I turn now to explicate the naturalistic credentials of the theory of systemic functions.

35. A similar point applies to the view of Bedau (1992) and the view of Enç and Adams (1992). Bedau appeals to the "good" or "value" of certain effects for certain agents. I do not dispute that some things are of value to various kinds of agents. On my view, however, while some systemic capacities are indeed good to or valued by some agent, that is not a requirement for functional status among natural traits. No doubt some systemic capacities are valued, but many are not valued by anyone, yet that does not prevent them from being functions. To think otherwise is, I suspect, to insist on the distinction between functions that are "proper" and effects that are merely useful. And that, as I have been arguing, is a distinction we ought to reject.

6

Guiding Inquiry: The Theoretical Roles of Systemic Functions

The final arbiter in philosophy is not how we think but what we do.
—Ian Hacking 1983

The theory of selected functions, as we have seen, aspires to explicate the concept of functions employed in modern biology. The aim is to specify historical conditions that biologists implicitly or explicitly have in mind when attributing functions. And according to the theory, the conditions biologists have in mind concern the selective success of functional traits; biologists attribute functions when they believe that the persistence of the functional type is caused by the selective success with which ancestral tokens performed the specified task. By contrast, the theory of systemic functions aspires to understand the concept of functions by considering the roles it plays in inquiry into hierarchical systems. The emphasis is on the theoretical work that function attributions actually accomplish rather than the way the concept is implicitly or explicitly understood. Emphasis is on what life scientists actually do and not on what we think they have in mind.

Systemic functions play several roles in inquiry. These include the historical explanations to which advocates of selected functions appeal. The theory of systemic functions, as explicated in chapter 3, attributes functions to traits the effects of which contributed to their own selective success and thus contributed to evolution or stasis in the population at large. The theory thus can explain everything explicable from within the theory of selected functions. But systemic functions have other important theoretical roles. Some of these roles depend upon the extent of our knowledge. We can explain, for example, how various lower-level capacities of known systems give rise to higher-level capacities, as the example of the circulatory system demonstrates. But when faced with unknown systems or parts of systems understood only partially, the role of systemic functions differs. Guiding, not explaining, is the focus. Systemic functions, when employed in systems we do not yet understand, guide us in our inquiries. They help us navigate the system by providing a provisional conceptual map, a provisional chunking of phenomena into tractable components. The ultimate goal, to be sure, is to explain how the system works, but that is accomplished only when we discover bottom-up mechanisms; the immediate goal is to provide entrée into the structure of the system. So the view that function attributions are explanatory, while true enough, is incomplete. The aim of this chapter is to consider the broader range of inquiry-guiding roles that our attributions of functions occupy, especially when faced with a system understood partially or not at all.

My thesis is twofold. I shall argue that the theory of systemic functions, in its various inquiry-guiding roles, offers a thoroughly naturalistic approach to understanding functions. This can be seen at two levels—the theory taken as a whole and the attribution of specific systemic functions. Specific attributions are naturalistic because the theory requires evidence for physical mechanisms that instantiate the attributed functions. The theory as a whole is naturalistic because it is an integral part of inquiry into hierarchical systems and because the ontology of the theory is thoroughly naturalistic. I shall also argue that, because the theory of systemic functions eschews properties that are abstract or noncausal, it cannot account for the occurrence of malfunctions as usually conceived. The alleged normative status of functional properties must be understood differently. To this end, I continue the Humean speculations begun in chapters 4 and 5. I suggest that a token trait that has lost the relevant systemic capacity no longer belongs to the associated functional kind, in which case it cannot be classified as malfunctioning. Nevertheless, given our knowledge of the larger system, we expect such tokens to exercise the relevant capacity. And when our expectations are thwarted by damage or disease, we are inclined to persist in seeing the token in terms of these expectations; we persist in placing the damaged and diseased in the functional type. That is why we are inclined to (mistakenly) claim of such tokens that they are malfunctional.

I The Inquiry-Guiding Role of Systemic Functions

We should, I believe, reject the general methodology employed by the theory of selected functions. As I argued in the previous chapter, we should reject the attempt to understand functions by examining the conditions under which modern biologists apply the term. Biologists are concerned to understand specific biological phenomena, but the task of integrating their concept of functions into the natural sciences generally is not a core concern. It is of concern, however, in the philosophy of science. The task is to discover whether or not we can understand functions in a way that coheres with the methods and postulations of the natural sciences. We should not ignore the way biologists employ the term or the concept, but we ought not assume that their usage is the result of careful reflection upon the philosophical question. We ought to approach the topic of functions differently.

I believe we do better to consider the theoretical roles that function attributions fulfill—the tasks accomplished by the attribution of functions when trying to render intelligible the workings of hierarchical phenomena. Instead of asking, Under what conditions do biologists attribute functions? consider the question, What theoretical *work* is accomplished by attributing functions to natural traits? or, What are we, *qua* inquirers, *doing* with our function attributions? Answering these questions, I suggest, is far more likely to lead to a satisfactory naturalistic account of functions. On this view, systemic functions are action guiding and, more specifically, inquiry guiding.

The prevalent view, by contrast, is that function attributions are explanatory. The theory of selected functions, in particular, asserts that the attribution of a function is equivalent to a selective explanation of the persistence of the functional type. I do not deny that function attributions are sometimes explanatory in this way, but I do not agree that this sort of explanation is the main theoretical role that our attributions of functions play. Instead, attributing functions often serves to guide our inquiry into hierarchical systems we do not yet understand well. Our attributions accomplish this in two general ways. First, the attribution of systemic functions provides a top-down taxonomy of a system that is highly provisional in character. This taxonomy does not explain, but rather provides a preliminary map with which to parse the system and study its functional parts. As we uncover increasingly specific systemic components, we revise or extend our initial taxonomies. And, second, until we discover quite specific physical mechanisms instantiating the systemic capacities postulated at the lower levels of analysis, we must refrain from claiming that our function attributions are secure. The systemic capacity analysis is top-down at the outset, but it cannot be accepted in full until its various attributions are tested from the bottom up. It is this that makes some function attributions explanatory: We can explain why a system behaves as it does once we have confirmed a taxonomy of functions by pointing to the systemic mechanisms that instantiate the relevant systemic functions. This is just to say that our explanations, insofar as they are mechanistic, require evidence of the mechanisms to which we appeal. On this view, then, function attributions are inquiry guiding before they are explanatory. They first provide a conceptual wedge into complex phenomena and a set of signposts to guide us toward the causally efficacious mechanisms of the system.

The entering point in any such investigation is the higher-level systemic capacity we wish to understand. Understanding the implementation of some such capacity is the goal of our analysis. It also marks the point of departure for our top-down attempt to taxonomize the system into salient capacities and components. For systems that are already understood to a significant degree, both the top-down taxonomy and the instantiating mechanisms are before us, in which case our attributions serve to explain how the system works, how it instantiates some higher-level capacity. Any system for which we can produce a mostly complete flowchart of its operations is one for which the inquiry-guiding role of systemic functions is greatly diminished while the explanatory role is greatly enhanced. But for systems we do not yet understand or understand only partially, our function attributions serve as a guide for increasingly detailed inquiry into the system. I will illustrate with two relatively simple examples, the first discussed by Bechtel and Richardson (1993) and the second by Enç (1979).

Consider the early investigations of the brain conducted by Franz Joseph Gall during the first half of the nineteenth century. Gall rejected the dominant view that the brain is a homogeneous unit resistant to mechanistic analysis. He hypothesized that, to the contrary, the brain is composed of several organs or centers, each responsible for a quite specific character trait. His aim was to provide a mechanistic analysis of various intellectual capacities by postulating a variety of brain organs. He postulated distinct cerebral organs for capacities such as speech, understanding time, navigating the environment, and so on. He also maintained that the various organs of the brain develop differently across individuals, growing larger in some and smaller in others, thus accounting for individual differences in the relevant capacities. The larger the neurological organ, the greater the correlated capacity. Assuming that the cranium is sufficiently malleable to show individual differences in the size and shape of these brain organs, Gall thought he could determine the degree to which an individual possessed a given character trait by examining portions of the cranium. Hence the practice of craniology.

Although Gall's phrenology came in for a great deal of criticism and derision, his basic strategy was commendable. Along with fellow phrenologist Johan Gaspar Spurzheim, Gall believed they had ample evidence that our psychological faculties were the products of our brain. Spurzheim called this the "First Principle of Phrenology." The brain, they assumed, had to have sufficient internal complexity to underwrite the wide variety of psychological capacities observed in human beings. Having thus rejected the view that the brain was unanalyzable, they considered the question, What are the functions of the brain such that human beings have the intellectual and practical capacities they in fact exhibit? This is to assume, as a provisional guide to further inquiry, that the brain is susceptible to a systemic function analysis. Now, the particular answers offered—that there is a neurological organ corresponding to each intellectual or practical capacity and that we can discover such organs via craniology—were mistaken, although a remnant of Gall's organology survives in contemporary cognitive psychology in the thesis of modularity. But the basic approach—analyzing the system into functional parts relative to some specified set of higher-level capacities—was clearly on track. Thus the inquiry-guiding power of systemic functions.

But this does not exhaust the role of systemic functions in this story. The empirical investigations conducted by phrenologists never matured. Gall began by developing a top-down taxonomy of neurological functions intended to account for intellectual and practical capacities. Unfortunately, the only confirmation for this taxonomy came in the form of correlations between features of the cranium and capacities of the intellect. The systemic function analysis failed to progress further. In particular, it failed to descend to the capacities and mechanisms within the brain itself. The claim that there existed a one-to-one correspondence between cranial characteristics and capacities of the intellect, even if confirmed by observation, does not tell us how the neurological system instantiates capacities of the intellect. Of course, the hypothesis that the brain is composed of distinct organs, each responsible for a specific capacity of the intellect, is a nontrivial hypothesis. It stakes a claim about the structure of our brains. But this claim was never put to the test by phrenologists; they failed to pursue the bottom-up portion of the theory of systemic functions. Had they done so, they would have come under strong pressure to relinquish their view. Indeed, once it was discovered that the brain is not organized in the ways described by phrenology, the proposed top-down taxonomy was no longer tenable. This demonstrates more vividly the inquiry-guiding role of systemic functions: The discovery that the postulated capacities are not instantiated in any systemic mechanisms provides powerful reasons for rejecting the proposed taxonomy and beginning the search for an alternative.[1]

Now consider a case in which our knowledge of the relevant system, while incomplete, was not nearly so immature. James Watson wanted to understand the synthesis of complex proteins. This process involved the rather precise ordering of different amino acids. Watson wanted to know the mechanisms responsible for it; he wanted to understand how the process worked. He reasoned that it could not be accomplished by special enzymes, since enzymes are themselves complex proteins formed by the very process in question. He thus hypothesized that the ordering was accomplished not by further enzymes but rather by some sort of template that had the capacity for self-replication. Watson put his suggestion this way:

> It is therefore necessary to throw out the idea of ordering proteins with enzymes and to predict instead the existence of specific surface, *the template,* that attracts

1. See Bechtel and Richardson (1993), chapter 4, for a fuller discussion.

> the amino acids (or their activated derivatives) and lines them up in the correct order. . . . It is furthermore necessary to assume that the template[s] must also have the capacity of serving, either directly or indirectly, as templates for themselves (self-duplication). (Watson 1970, 179–80, quoted in Enç 1979, 352)

As Enç stresses, Watson's appeal to a template is clearly intended to refer to some mechanism, as yet undiscovered, that instantiates the function of ordering amino acids. Unlike the situation facing Gall, Watson had ample background knowledge from which to reason. He could say quite specifically what the template could not be: It could not be an enzyme. He could also say that the template had to be some feature within the cell endowed with the capacity for self-duplication. All of this is to say that the postulation of a template—a systemic capacity of cells—played a crucial role in the discovery of messenger RNA. The attribution of a systemic function guided the search for the RNA template. It provided a piece of the taxonomic puzzle that then imposed constraints on the sorts of mechanisms for which to search.[2]

Enç (1979) discusses other examples—including Harvey's discovery of the function of the heart—that illustrate the inquiry-guiding role of systemic functions. He does not put the point in terms of guiding inquiry; in fact, he asserts that function attributions are explanatory. But it is clear enough from the case of messenger RNA that "explanatory" covers a broad range of theoretical contexts, including those that serve to guide the process of inquiry rather than explain the occurrence of certain events. Moreover, Bechtel and Richardson (1993) develop a wealth of splendid case studies that illustrate the inquiry-guiding role of systemic functions in detail. Now, Bechtel and Richardson are not concerned to defend any particular theory of functions. They are concerned, rather, to defend the claim that progress in scientific discovery often involves the application of at least two heuristic strategies, namely, *decomposition* and *localization.* But these heuristics overlap the strategy of the theory of systemic functions in significant ways. In consequence, the several case studies developed in Bechtel and Richardson serve as evidence for my claim that systemic functions play a central role in guiding inquiry into complex hierarchical systems.

2. For fuller discussion of this case see Enç (1979), 352–55.

To see the overlap between Bechtel and Richardson's heuristics and the theory of systemic functions, consider first the process of decomposition. Their account of this process is sophisticated and nuanced, but the basic steps involve the delineation of a system (or what we tentatively take to be a system), identification of higher-level capacities we wish to understand, and analysis of the system into functional components that together produce the higher-level capacities. Like analysis in terms of systemic functions, the process of decomposition is recursive, proceeding as far as our interests drive us or as far as systemic complexity admits. The process of decomposition thus applies only to systems that are hierarchically organized—that is, systems that can be analyzed into distinct levels of organization. The ultimate goal of this top-down strategy is discovery of physical mechanisms internal to the system responsible for the functional capacities listed in our taxonomy. This involves the second heuristic, localization. It is assumed that the capacities of interest are instantiated by mechanisms within the system. They need not be discrete or otherwise tidy; they may be distributed throughout the system. But until we uncover substantial evidence that there are physical mechanisms the effects of which give rise to the identified capacities, we must temper the confidence with which we accept our functional taxonomy.

The strategies of decomposition and localization, or close cousins of these heuristics, are part of the theory of systemic functions as developed by Cummins (1983) and as developed in chapter 4 of this discussion. Application of the theory of systemic functions requires identification of a system, a systemic capacity we wish to understand, and an analysis of the system into functional capacities at two or more levels of organization. This is Cummins's *analytical strategy*. Its similarities to the process of decomposition are obvious.[3] Cummins further describes what he calls the *instantiation strategy*. It involves tracing the links of causation to the physical mechanisms operating at lower levels of organization. Instantiation and localization are close cousins indeed, for both require that the

3. Cummins does not restrict application of the theory to hierarchical systems, at least not explicitly. That restriction is unique to my formulation of the theory, though inspired by Bechtel and Richardson (1993).

top-down strategy be confirmed by the bottom-up identification of causally responsible mechanisms. The importance of bottom-up confirmation is emphasized by Bechtel and Richardson throughout their discussion, including the following:

> Showing how systemic functions are, or at least could be, a consequence of subtasks [those specified in the decomposition] is an important element in a fully mechanistic explanation. Confirming that the components realize those functions [by way of localization] is also critical. Both are necessary for a sound mechanistic explanation. (Bechtel and Richardson 1993, 125)

And as we saw in chapter 2, Cummins insists that "[f]unctional analysis of a capacity C of a system S must eventually terminate in dispositions whose instantiations [in physical systemic mechanisms] are explicable via analysis of S. Failing this, we have no reason to suppose we have analyzed C as it is instantiated in S" (Cummins 1983, 31).

That said, there are differences between the two views; I discuss one such difference in the next section. But in the context of defending the theory of systemic functions, the similarities outweigh the differences. The case studies offered by Bechtel and Richardson thus provide powerful evidence of the inquiry-guiding role of systemic functions. All of their studies trace the dynamic processes involved in scientific discovery. They are concerned to explicate psychologically realistic strategies with which we make scientific discoveries. They focus upon complex and hierarchical systems which, at the relevant historical juncture, were understood partially or not at all. Their claim is that, among the various strategies we bring to inquiry into such systems, decomposition and localization are of great utility. Their argument for this claim consists in an examination of several historical cases of scientific discovery, including discovery of the mechanisms involved in fermentation, respiration, language production and comprehension, spatial memory, developmental genetics, and more. In each case, the attribution of systemic functions plays the role of providing a conceptual roadmap, directing our attention to the likely sorts of mechanisms that instantiate the capacities we wish to understand. And in each case, the confidence with which we attribute such functions increases as we confirm, from the bottom up, the existence of the requisite sorts of mechanisms. Hence the naturalistic status of specific systemic function attributions. Moreover, the theory of systemic functions, taken as a whole, is integral to the actual strategies with which we study hierarchical systems. Hence the naturalistic credentials of the theory.

There are cases, of course, in which the discovery of lower-level physical mechanisms does not provide a complete explanation of the higher-level capacity under study. Churchland and Sejnowski (1992), drawing on Selverston (1988), offer a nice example. Selverston studied the stomatogastric ganglion of the spiny lobster—a twenty-eight–neuron circuit responsible for the muscles involved in grinding food. Despite having described the anatomical and electrical properties of each neuron, Selverston could not explain the rhythm of the circuit's outputs. The grinding behavior of the lobster is explained by the fact that the neurons fire in a given rhythm. What is not clear, however, is how the circuit produces its own rhythm. Pointing to the causal powers of the neurons is not enough, for no one cell (and presumably no cluster of cells) produces the rhythm. Churchland and Sejnowski thus suggest that, in addition to the powers of the mechanisms, we also need to discover the "interactive principles" that govern the system (Churchland and Sejnowski 1992, 5).

Now, I do not think we should dismiss the possibility of such principles out of hand, but nor do I think that we should accept them on their face. Two points are relevant. First, we should expect that any such interactive principles will be, upon further investigation, explicated in terms of *some* physical mechanisms. The lobster's pattern of grinding may not be explicable in terms of any one cluster of neuronal cells, but surely there exists an explanation—one that appeals to *some* physical properties of the organism or the organism's environment—that accounts for the pattern. *Something* in the constitution of the organism or its physical surroundings accounts for the fact that it grinds in *this* pattern rather than some other. To this extent, appeals to interactive principles are appeals to ignorance; they are principles waiting for a fuller mechanistic explanation; indeed, they are appeals to top-down systemic functions waiting a complete analysis. Second, we should, I suppose, leave open the possibility that mechanistic explanations are limited in scope and that some systemic capacities cannot be adequately explained by appeal to physical mechanisms. This strains credulity, of course, given the stunning success of mechanistic explanations to date. Nevertheless, naturalism, as I construe it, is a commitment not to a specific ontology but rather to the methods of inquiry employed by our best natural sciences. So, if nonmechanistic explanations are clearly successful in some domain, and if such explanations appeal to nonmechanistic principles of interaction, then so be it—we should embrace the reality of such principles. And in that event, the theory of systemic functions would retain its naturalistic credentials by appealing to principles of interaction in addition to physical mechanisms.

The inquiry-guiding role of systemic functions contrasts sharply with the explanatory import assigned to selected functions. The central thesis in Wright (1973) is that the attribution of an etiological function is equivalent to a historical explanation of the persistence of the functional type. As we saw in chapter 5, advocates of selected functions reject Wright's broad notion of selection and restrict their explanations to natural-selective explanations within evolutionary theory. And in light of this restriction, the theory of selected functions is essentially explanatory and historical. By contrast, the theory of systemic functions allows that the attribution of functions is sometimes explanatory on historical grounds, other times explanatory on nonhistorical grounds (Harvey on the function of the heart), and, most notably, sometimes not explanatory at all. As Enç (1979) and Bechtel and Richardson (1993) demonstrate, inquiry into hierarchical systems that are not yet understood involves the attribution of systemic functions, the role of which is nonexplanatory. Or, at minimum, explaining is not the immediate role of such attributions. Their immediate role is to provide a tentative taxonomy of systemic capacities, that we might formulate hypotheses concerning the sorts of mechanisms instantiating those capacities. We want to locate and describe the causal mechanisms that form the system, that we may predict and explain how the system does what it does. The appeal to systemic functions helps us explain, to be sure, but only after the system is understood; before that, the appeal to systemic functions guides but does not yet explain.

Some may wish to challenge the distinction I have drawn between explaining and guiding inquiry. After all, attributing systemic functions in systems we do not yet fully understand plays an explanatory role. It tells us how the system might be working; it provides a how-possibly explanation of the specified systemic capacity. I agree that the attribution of systemic functions can be construed in this way. But the distinction between explaining and guiding inquiry is important nonetheless. It is true enough that, in attributing systemic functions to systems poorly understood, we venture a kind of how-possibly explanation.[4] The relevant question, however, is, What *role* do these sorts of function attributions play in our inquiry? My answer is that that they guide inquiry into the unknown components of the system. They provide a provisional mapping of the system that we then proceed to test. That is the payoff of such attributions; that is *why* we make them. And these attributions do not become genuinely explanatory—they do not acquire the role of providing an actual *account* of the way the system works—until we have discovered evidence that the attributed functions are instantiated by mechanisms in the system. Systemic functions in systems poorly understood are best construed not as explanations but rather as potential explanations, the immediate goal of which is to guide us into the system's structures.

The requirement that we locate physical mechanisms responsible for instantiating the systemic capacities itemized in our taxonomy ensures that the attribution of functions is well confirmed. When functions are attributed to systems well understood, such as the circulatory system, we have extensive knowledge of the physical mechanisms that instantiate the larger systemic capacities. Our function attributions, in this case, ought to engender confidence. When attributed to systems poorly understood, our function attributions are tentative and hypothetical. Their existential import depends upon the degree to which available evidence confirms their existence in the system. In the early stages of inquiry into a given system, our attributions are highly tentative and our confidence must be tempered. We must await bottom-up confirmation before asserting that the functions in our proposed taxonomy in fact exist. Our function attributions, in this case, are hypotheses with evident heuristic

4. There are two importantly different kinds of how-possibly explanation—two kinds of ignorance that they are intended to alleviate. One kind is given when we are ignorant of the specific processes involved but not of the general structure of the larger system. Kingsolver and Koehl's (1985) explanation of the evolution of insect wings, discussed in chapter 3, is illustrative. The structure of wings and indeed the entire insect was known from fossils. But another kind of how-possibly explanation is given when we are ignorant of the overall structure of the phenomena we are trying to understand. Gall on the brain and Watson on RNA are illustrative. And here the guiding role of systemic functions is most pronounced.

value, but hypotheses to be tested all the same. Such attributions acquire their naturalistic credentials on the grounds that they require empirical confirmation.

II The Limits of Decomposition and Localization?

Despite my appeal to Bechtel and Richardson (1993), there is one aspect of their view I wish to question. Toward the end of their discussion they claim that there are certain systems to which decomposition and localization do not apply. Some systems exercise higher-level capacities by virtue of highly integrated interactions among lower-level capacities. A high degree of interaction means that the various lower-level components cannot produce their salient causal effects independent of the effects of most or all other components. Significant integration among systemic components excludes significant independence of operations. The problem, according to Bechtel and Richardson, is that, as integration increases and independence decreases, the system becomes increasingly less decomposable. Decomposition assumes some minimal degree of causal independence among some or most components; we cannot assign specific systemic functions to any component unless we can discover what its unique causal contribution to the system is. But if the system is highly integrated, we are barred from any such discovery because we cannot isolate the efficacy of any one component. So, decomposition fails. And if the system cannot be decomposed, localization cannot even begin. Decomposition and localization, therefore, are limited to systems in which the degree of integration remains below a certain threshold. If, as I have suggested, the theory of systemic functions substantially overlaps these heuristics, then the applicability of systemic functions is similarly limited. It is this claim I wish to question.

Bechtel and Richardson describe two connectionist models. These, they suggest, exemplify the sorts of integration that render decomposition ineffectual. The first model is a simple two-layered network that, after fifty epochs (training runs), acquires the capacity to recognize and classify cups, buckets, hats, and shoes. More modestly, the system acquires a capacity that can be interpreted in semantic terms as involving the capacity for recognition. The second model is a five-layered network, developed by Hinton (1986), that acquires the capacity to recognize and classify the members of two family trees after 1500 epochs. The details of both cases are interesting but not essential to understanding the central claim. The result of training is that certain patterns of activation between various processing units (nodes) are correlated with certain features of the observed objects or relations. After training, when certain features are presented to the network, the associated activation pattern tends to fire. It thus is plausible to interpret these activation patterns as having semantic content, as referring to or being about the correlated feature of the object presented.

The problem, however, is that the recognitional capacity of the network, though composed of patterns of activation distributed across several nodes, is not decomposable into meaningful subtasks. The lower-level interactions are far too integrated to be appropriately decomposed. Bechtel and Richardson conclude as follows:

> It is not important that we take a stand on the ultimate viability of connectionism as a framework for cognitive theorizing in order to make our main point: connectionism represents a break with traditional mechanism, pointing toward a different category of models and employing an alternative strategy for developing them. This alternative emphasizes systems whose dynamic behavior corresponds to the activity we want to explain, but in which the components of the system do not perform recognizable subtasks of the overall task. . . . The overall architecture of the system—and especially the way the components are connected—is what explains cognitive capacities, and not the specific tasks performed by the components. We have abandoned decomposition and localization. (Bechtel and Richardson 1993, 222–23)

If we have abandoned these heuristics, we have likewise abandoned an analysis of the system into systemic functional components. A few pages earlier they summarize their claim in similar terms:

> Network models do account for the cognitive performance, but often they do so without providing an explanation of component operations that is intelligible in terms of the overall task being performed. The network is a cognitive system; the components are not. The result is that we do not explain how the overall system achieves it *[sic]* performance by decomposing the overall task into subtasks, or by localizing cognitive subtasks. (214)

I agree with most of the premises in these arguments, but not all, and I disagree with the conclusions. As I read it, the above objection is, at bottom, a worry about the limits of our knowledge and perhaps our understanding of our own models. The worry is about the extent of our present ignorance. Worrisome as our ignorance may be, it does not, in my view, warrant the pessimistic claim that decomposition and localization cannot be applied. At least three considerations are relevant.

(1) I agree with Bechtel and Richardson that the "overall architecture of the system—and especially the way the components are connected—is what explains cognitive capacities, not the specific tasks performed by the components." I agree, that is, that the connectionist model accounts for the higher-level capacity (recognition) in terms of the *structure* of the network—in terms of patterns of activation—and not in terms of specific *tasks* performed by specific lower-level devices. But we should not agree that, in consequence, a systemic function analysis cannot be given. For we sometimes explain the instantiation of a higher-level capacity by analyzing that capacity into the organized structures that constitute the system at its lower levels. The suggestion is that we conceptualize connectionist models, at least at the outset, as offering analyses of cognitive capacities in terms of structural—not interactive—systemic functions. Just as Haugeland (1978) analyzes the capacity of a fiber-optic cable to transmit images into the structural features of the cable itself—the size, location, and number of individual fibers—so too we can analyze the capacity of a connectionist network to exercise some recognitional capacity in terms of various structural features of the network—patterns of activation and weighted connections. This seems to me an acceptable, even if simplistic, application of the theory of systemic functions.

On the account of decomposition offered by Bechtel and Richardson, higher-level systemic capacities are properly analyzed not into lower-level structures, but rather into lower-level tasks, and this seems to me an overly restrictive notion of decomposition. After all, the ultimate goal of decomposition is the localization of lower-level mechanisms responsible for the higher-level capacity. And this is accomplished often enough by analyzing the system into lower-level structures. We certainly accomplish this in our analysis of Haugeland's cable. And, as Bechtel and Richardson concede, connectionist models do indeed explain cognitive capacities by appeal to the "overall architecture of the system" rather than the specific tasks performed by systemic components. Why, then, insist that decomposition must appeal to lower-level tasks and not to lower-level structures?

Bechtel and Richardson hold that a system can be decomposed only when there is some degree of causal independence among lower-level components, and that seems right. The question, then, is whether or not lower-level structures exhibit the relevant sort of causal independence; the question is whether or not we can, by experimental means, isolate the causal contribution of a given structure to the more general systemic capacity. It seems to me that we can in the case of Haugeland's fiber-optic cable. We can measure quite precisely the quantity and wavelength of light conducted by individual fibers. Likewise for the connectionist models described above. We can identify quite specific patterns of activation that fire when the system is presented with appropriate input. Indeed, it is the correlation between specific activation patterns and the presence of certain objects that underwrites our claim that the patterns have some sort of semantic value. We ought to insist, therefore, on a notion of decomposition that appeals to the organized causal effects of lower-level structures.

(2) There may be a quite different worry driving Bechtel and Richardson's claim concerning the limits of decomposition and localization. On the analysis given so far, we analyze the semantic capacity of a connectionist system into lower-level structural components, specifically, into lower-level patterns of activation. More to the point, we analyze a single cognitive capacity into a single pattern of activation. This, however, may jar our sensibilities. We may find it unsatisfying, indeed puzzling, to claim that a given semantic capacity is *explained* simply by identifying the pattern of activation involved. Such an analysis moves immediately from a higher-level cognitive capacity (recognition) to a lower-level structure (pattern) that is difficult to interpret semantically. There is a sudden shift from the cognitive to the noncognitive and perhaps this is the real difficulty in trying to decompose connectionist systems. Indeed, at times Bechtel and Richardson seem to assert that connectionist systems are not decomposable at all (full stop), but other times they suggest that these systems are not decomposable into component capacities that are "intelligible in terms of the overall task being performed." The result, they say, "is that we do not explain how the overall system achieves it *[sic]* performance by decomposing the overall task into subtasks, or by localizing *cognitive* subtasks" (1993, 214, my emphasis). This seems to suggest that

the real worry is the abrupt move from a higher-level cognitive capacity to lower-level capacities that are barely cognitive or completely noncognitive. It is as if we can see the structures involved in the production of the higher-level capacity but have no intuitive understanding of how or why this capacity emerges out of these structures. We have the feeling that something important—namely, our understanding—has been left out. At the close of their chapter on highly integrated systems, and referring specifically to connectionist networks, they say: "We may not be able to follow the processes through the multitude of connections in a more complex system, or to see how they give rise to the behavior of the system. We may fail in the attempt to *understand* such systems *in an intuitive way*" (229, my emphasis).

If this is the root of their complaint, then there is no especial objection to the applicability of the theory of systemic functions or of decomposition and localization, for the sense we have that something is left out is not unique to connectionist models of cognition. The very same problem afflicts traditional computational models as well—though it tends to arise later in the analysis of a system. Cummins (1983), for example, worries over this (and a related) problem.[5] Traditional computational models analyze complex cognitive capacities into less-complex cognitive capacities interacting at some lower level of organization. This strategic step is iterated for each level at which some form of cognitive activity occurs. Finally, however, our analysis of low-grade cognition takes us to the level of capacities and components that are entirely mechanical. It is at this point that the above worry arises. Our analysis of some lower-level cognitive capacity in terms of purely mechanical components may give the sense that something is left out. We may have the feeling that we do not understand how or why these specific mechanisms, which by hypothesis are devoid of cognitive capacities, give rise to cognitive capacities at the next level up.

Of course the gap between the cognitive and the noncognitive is greater in connectionist models, since we move quickly from the fully cognitive to the noncognitive or, at a minimum, to a level of nodes or sets of nodes that are difficult to track in semantic terms. Perhaps this explains why

5. See Cummins (1983), chapter 2.

some theorists find connectionist models ultimately unsatisfying—they appear to leave out more than traditional computational models. But the crucial point is that decomposition and localization apply without problem to traditional computational models of cognition, as does the theory of systemic functions. So the difficulty involved in bridging the cognitive/noncognitive gap provides no particular reason for thinking that these heuristics or this theory cannot apply to connectionist models as well. If we have the sense that rather little fruit is to be harvested from analyzing connectionist systems in terms of systemic functions, we should suspect that this is a deficiency in the connectionist network under consideration. Or perhaps we should suspect that our inquiry into the capacities and mechanisms at lower levels of neurological organization is too undeveloped for us to assess the functional taxonomy that our connectionist models provide.

(3) Of course Bechtel and Richardson may be right that there are some hierarchical systems to which the theory of systemic functions does not apply. Perhaps there are systems constituted not out of layers of structural and interactive mechanisms, but units of some other type. Nevertheless, I am not convinced that connectionist models are an instance of this unfamiliar type. Inquiry into the neurological basis of our cognitive capacities is still maturing and it is hard to predict what advances technology will bring. We know too little about the neurological mechanisms that underwrite, or may underwrite, our most accurate connectionist models. On my view, the analysis of a cognitive capacity into a pattern of activation is merely the first step. It remains for us to investigate further the neurological mechanisms involved in the production of such patterns. And as I have been urging in this chapter, the attribution of systemic functions often occurs in trying to understand a system poorly or partially understood. Such attributions guide more than they explain. We should understand connectionist models in this way. Our knowledge of the implementation of connectionist models depends on progress in neuroscience, and while great strides have been made in recent years, much about the brain remains unknown. To the extent we can attribute systemic functions to connectionist components—even if only at the level of activation patterns—we do so as a guide to further systemic inquiries, including inquiries into the functional capacities of the brain. So,

I agree that we should be cautious in the attributions we make, but I do not agree that we know enough to conclude that the heuristics of decomposition and localization are as limited as Bechtel and Richardson suggest.

Finally, even if Bechtel and Richardson are right about the limits of decomposition and localization, systemic functions nevertheless are of great utility in inquiry. And it is the depth and breadth of their inquiry-guiding roles that gives systemic functions both their scientific value and their sound naturalistic credentials.

III Systemic Malfunctions and Expectations

The attribution of systemic functions, then, is sometimes explanatory, while other times it serves to guide our inquiry into unknown phenomena. But this does not exhaust the theoretical roles of systemic functions. In attributing systemic functions, and in searching for the mechanisms that instantiate these functions, we engage in a process of classifying systems and their parts. We classify components according to the work they perform. Enç (1979), after noting the classificatory power of function attributions, argues that functional types impose limitations on the kinds of physical mechanisms in which such functions can be implemented. This is an important thesis—especially in philosophy of mind[6]—but I wish to draw a distinct lesson concerning the classificatory power of systemic functions. My suggestion is that the classificatory practice involved in applying the theory produces in us certain expectations concerning systemic types and tokens of those types. Indeed, these expectations inform our predictions and our explanations concerning the behavior of token traits. They thus facilitate our inquiries. But it is also these expectations that explain why we are inclined to say of certain token traits—those unable to perform the associated functional task—that they are malfunctioning. Or so I wish to suggest.

6. See Enç (1979), 349ff. Enç's argument appeals to features of natural kinds. In order for his claim to apply to traits such as hearts and eyes, however, we need an account of natural kinds that appeals not to microstructural properties but to something else, for otherwise it is hard to see how it applies to lineages, populations, species, or the like.

My aim here is to explain why we are inclined to classify some tokens as malfunctional. The aim is to account for the presence and efficacy of an inclination exercised in the course of inquiry. I am not interested in trying to explain how or why natural traits in fact malfunction, because on my view natural traits cannot malfunction. Of course, sometimes traits are defective; more precisely, sometimes we are inclined to conceptualize a token trait as defective. We are so inclined to the extent the token fails to perform in ways we expect. But, of course, the failure to perform in ways that satisfy our inclinations is not enough to show that the token is malfunctioning. And, as we have seen, systemic functions are identical to systemic capacities, in which case the loss of the relevant capacity entails loss of the systemic function. So traits that fail to perform in ways that match our expectations—expectations concerning the workings of the systemic type; more on this presently—are not malfunctional at all. They are merely nonfunctional. This is a consequence of the theory of systemic functions that cannot be dodged.

The basic point here is quite general: Any attempt to explicate functions in terms of some type of success—success in contributing to higher-level systemic capacities, for example—cannot account for the occurrence of malfunctions. This point applies to the theory of selected functions insofar as it defines functional types in terms of success in selection. The problem is that, if functional types are defined by reference to some form of success, then incapacitated tokens that lack the success property do not belong to the relevant functional types. Now, the aim of chapter 7 is to develop this argument against the historical approach generally; I defend the thesis that selected and strong etiological malfunctions are impossible. For present purposes, however, I assume that no theory of functions, including the theory of systemic functions, can account for the occurrence of malfunctions by defining functional types in terms of some sort of success. Incapacitated tokens thereby lack the associated systemic function. What remains, then, is to explain why we are so inclined to see nonfunctional tokens as malfunctional, why we insist on placing tokens that have lost the requisite physical capacity into their former functional categories.

I speculate that we are inclined to say of a nonfunctioning token that it is malfunctioning—that it should be performing such-and-such task even when it cannot—because we have acquired the expectation that components situated in systems of this type perform the stated task. Our expectations are caused by our knowledge of the type of system, or knowledge of analogous systems, and the workings of their parts. Our expectations are caused, in short, by prior applications of the theory of systemic functions. More specifically, our knowledge of complex, hierarchical systems causes us to expect the appearance of various component types, as well as a range of regular systemic effects produced by tokens of those types—that is, effects that contribute to the exercise of some higher-level capacity. We thus come to conceptualize components of the system in terms of systemic capacities that qualify as systemic functions. And when we happen upon a component token that fails to produce the systemic effects we have come to expect, we are inclined to (mistakenly) place the token into the relevant systemic functional type.

Consider, for example, an infant born with only one eye, where the other orbit contains a gelatinous blob that lacks the capacity to process light. We would be inclined to say of the blob that it is a malfunctioning eye or, at minimum, a malformed eye. After all, we know that human infants typically are born with an eye on either side of the face and we know a good bit about the developmental processes involved. On the basis of this knowledge, we expect the formation of two eyes endowed with the capacity to process light. When things do not go as we expect—when, for example, we happen upon the infant's blob—we are taken aback. We have the feeling that this token ought not be of this form; we have the feeling that this token has failed in some respect—that it has failed to become what it is supposed to be or that it is failing to do what it is supposed to do. But we are misled by these inclinations and, perhaps, by certain residual beliefs that, upon reflection, we would reject. The feeling we have that this token ought to be formed differently is an expression of our expectations concerning token components within this type of system—expectations, for example, concerning the systemic capacities of globular capsules crucial to the general capacities of the visual system. Precisely these sorts of expectations, I submit, are at the heart of our inclination to mistakenly see nonfunctional tokens as malfunctional.[7]

7. "Expectation" should not be construed narrowly. Some expectations depend upon knowledge of a given system acquired via repeated interactions with the system. But others may depend more on knowledge hardwired into our brains

Crucially, then, when we assert of some natural trait that it is supposed to perform some task, we are not attributing to that object a norm of performance. We are simply expressing our expectations of the token in light of our knowledge of the larger system. We thus are quite mistaken to describe the infant's blob as malfunctioning. We may also be wrong at a more basic level. We may be wrong even to classify the infant's blob as an eye. That is, membership in the category of eyes plausibly requires possession of the relevant systemic functions (as opposed to mere historical relatedness) and hence possession of the requisite physical capacity. The infant's blob, however, has no such capacity. So, it may be a mistake to describe the blob as a malfunctioning eye for the more elementary reason that it may be a mistake to place it in the category of eyes. We should not conclude, however, that there are no norms at play here. There are no normative properties ascribable to token traits, to be sure. Hence, there is no justified claim to the effect that this or that token occupies an abstract, noncausal role. Nevertheless, our expectations underwrite various epistemic norms. The basic standard is given by our knowledge of the larger system and the expectations this knowledge generates. We use this standard to formulate predictions and explanations of systemic behavior.

Consider, again, the speculations offered in chapters 4 and 5. Let us assume that natural phenomena are hierarchically organized and that, as hierarchical systems increase in complexity, the tendency to persist and perpetuate their kind increases. Add the further assumption that we are predisposed to want to understand how hierarchical systems work, including systems that are complex and persistent, and that we are predisposed to approach such systems by applying the theory of systemic functions. We are disposed, that is, to conceptualize systems in terms of

---

and less on experience of the system. Some expectations—those concerning faces and eyes, for example—probably include a deeply visceral component. A malformed eye grabs our attention and holds it, thanks to an innate capacity to pay attention to anything resembling a human face. This is not true of our expectations concerning natural systemic components generally. Surely some such expectations rest squarely on the results of inquiry into the relevant system. And, of course, we harbor expectations of the components of artifactual systems. Those expectations may be derived from experience of the system, but their main sources include various intentions and conventions.

higher-level capacities and the lower-level capacities in which they are implemented. With these assumptions in place, it is plausible that inquiry into complex hierarchical systems involves the categorizing of systemic components by reference to systemic capacities, by reference to contributions to higher-level capacities. And it is precisely here, in the process of classifying on the basis of systemic capacities, that we come to harbor expectations concerning the future form and effects of systemic components. These expectations form the basis for further inquiry into these and analogous systems, and the basis for our attempts to control and manipulate such systems for practical ends.

These expectations should be only as strong as our knowledge of the system warrants. In systems that we understand very little, we are prevented from seeing token components in light of informed expectations. We thus have little inclination to attribute malfunctions. In systems understood in substantial ways, the inclination to attribute malfunctions is proportionately greater. Our expectations are most potent for systems the workings of which we know well. The infant's visual system is a clear case in point. Our knowledge of human anatomy, of the mechanisms of inheritance, and the mechanisms of development all incline us to expect fairly specific systemic forms and capacities. But, of course, various events can result in relative formlessness or relative inefficacy, and here the feeling that something has gone wrong is the strongest. That is not surprising, given the extent of our knowledge of the visual system and our impressive sensitivity in perceiving faces.

As in chapter 5, I am in no position to defend the psychological conjectures to which I appeal; that is a job for psychologists. For purposes of this discussion, however, the essential point is the plausibility of the Humean strategy generally and the revision in our concept of functions that it entails. As suggested, there are two powerful reasons to forego the attribution of malfunctions and consider instead the hypothesis that incapacitated traits are nonfunctional rather than malfunctional. The first reason rests on the results of chapter 5. The alleged naturalistic standing of so-called "proper" functions is difficult to substantiate. The second reasons rests on the results of chapter 7. As I argue there, selected malfunctions are not possible. If we take both reasons to heart, we must revise our concept of functions, at least to some extent. Some theorists—Millikan (1989), Neander (1991), Preston (1998), etc.—insist that our concept of functions is a concept of properties that are inherently normative. My view, by contrast, is that, although there are good historical explanations for why our concept includes a normative component, there are no good philosophical grounds for retaining this component. On the contrary, to the extent we are naturalistically inclined, there are simple but sound philosophical arguments for rejecting any such component. Functions are systemic capacities, nothing more. Of course, there are many different kinds of systems and many different kinds of systemic capacities. The capacities of components within highly complex systems exhibit substantial regularity and tend to recur over time; the capacities of components within relatively simple systems may not. But what marks the difference between the complex and simple is not a difference in the kind of functional property at play, but simply a difference in the kind of system involved and, in some cases at least, a corresponding difference in the ways we are affected by the system.

The Humean strategy is also warranted on the grounds that it makes sense from within the theory of systemic functions as a whole. As we have seen, the expectations that underwrite our inclination to (mistakenly) attribute malfunctions arise when we employ the theory of systemic functions. These expectations are the product of conceptualizing systemic components in terms of how they work within the larger system; they are the product of identifying from the top down and confirming from the bottom up the existence of systemic functional types within a system. In the ideal case, such expectations are the fruits of our theoretical labors. They also guide us in those labors. They enable us to anticipate, predict, control, and render intelligible the behavior of tokens of these types. In some cases, however, these expectations mistakenly lead us to see nonfunctional tokens as possessed of properties they do not have. The mistake comes in thinking that there are norms that attach to natural traits, when in fact the operative norms are nothing more than the expectations with which we approach various natural systems. This way of understanding our inclination to attribute malfunctions is recommended, then, by placing the practice of producing such attributions within applications of the theory of systemic functions.

Finally, this approach to the topic of malfunctions taps into a key insight of the view defended by Bigelow and Pargetter (1987)—the selective propensity view.[8] This view has been criticized in various ways, but the basic insight that function attributions are in some way forward-looking still stands. The theory of systemic functions, as developed in this book, captures the forward-looking feature of functional properties and does so in ways that make it superior to Bigelow and Pargetter's own account. Two considerations are relevant.

First, recall that, on their view, traits possess functions insofar as they have the propensity to be selected for the performance of certain tasks. Functions are dispositional properties of traits, properties that would be selected for were the organism's natural habitat to remain the same. I accept part of this view. I accept that functions are capacities belonging to systemic components and that these capacities need not be exercised in order to qualify as functions. The capacities must in fact belong to the systemic component; indeed, we must have evidence of physical mechanisms capable of instantiating the capacity within the system. But conditions external to the system may be such that the capacity is never exercised, that the underlying mechanism is never activated, in which case the functional capacity is idle but present nonetheless. But there is also part of this view I reject. Bigelow and Pargetter hold that functions are capacities that will be selected for under relevant conditions; this is to restrict the realm of functions to the potential for success in selection. This I do not accept. On my view, functions are the capacities of components that give rise to the exercise of higher-level systemic capacities. Whether or not these capacities have the potential to be favored by selection is quite beside the point—except in cases where the higher-level capacity we wish to explain (a population's capacity to evolve via selection, for example) is instantiated in mechanisms of selection. The internal economy of organisms can continue its various tasks—circulation, respiration, digestion, etc.—even while being selected against. Systemic functions persist and sometimes even flourish in the face of negative selective forces. And if the world were such that selection never occurred but there nevertheless existed hierarchical systems with structural and interactive components operating at lower levels of organization, the components of these systems would possess a range of systemic functions. To fixate on the functions that emerge (or would emerge) in the process of selection

8. Explicated briefly in chapter 2, section III.

is to fixate on one among several general kinds of systemic functions (see chapter 3 for further discussion).

The second consideration is that the forward-looking feature of functions is more compelling on my view than on the view of Bigelow and Pargetter. On their view, functions are forward-looking just because they are properties that will be selected for if the environment remains stable in salient ways. The forward-looking aspect of functions is a feature—a subjunctive property—of the functional dispositions themselves. The dispositions look forward in the sense that, if the habitat of the organism does not change significantly, the dispositions will be selectively successful in the future. But this, I believe, misses the key insight, for it seems highly likely that there are dispositions that meet Bigelow and Pargetter's subjunctive conditions, and hence qualify as functions, that nevertheless are not forward-looking in their sense of the term. To see this, consider that, on their view, there is an indefinite number of functional properties that exist unmanifested. There are dispositions belonging to organisms that, because of some change in habitat, have not been selected for. And it seems safe to assume that there also are dispositions that *never* will be selected for because of irreversible changes in habitat. We may not know which dispositions these are, but it is plausible that numerous such capacities exist. And it is hard to see in what sense these unmanifested dispositions are forward-looking. If the probability of their ever being selected for is practically zero, then there is nothing of the future that they will ever see, in which case they are not forward-looking in Bigelow and Pargetter's sense.

To capture the insight that functions are forward-looking, we should focus not on the propensity for selection but rather on the practice of attributing functions. The central thought is that functions are attributed with an eye toward the future behavior of the functional trait. They are attributed, more specifically, on the basis of our informed expectations derived from our study of the relevant type of system. They are attributed insofar as application of the theory of systemic functions has generated in us the appropriate expectations. So, the forward-looking aspect of functions is not a feature of the systemic capacities, as Bigelow and Pargetter suggest, but rather an epistemic feature of the attribution of systemic functions. It is *we*—not the systemic capacities—who look forward in the course of attributing systemic functions. We look to the future insofar as we expect future tokens of the systemic functional type to possess and to exercise the relevant systemic capacity. And, while there is nothing intrinsically forward-looking about functions, the forward-looking aspect of function attributions is of great importance nonetheless. It is important in scientific inquiry. It is part and parcel of the inductive inferences we make in the course of such attributions. It is also why we are so inclined to (mistakenly) say of tokens that disappoint our expectations that they are malfunctioning.

Systemic functions thus play a variety of roles in the course of inquiry. When attributed to the components of poorly understood systems, our attributions are tentative and hypothetical with little or no existential commitment. When attributed to systems that we understand well, our attributions provide an account of the way the system works. This account forms the basis for our explanations and predictions regarding such systems. Our attributions also express our informed expectations regarding the functional types and these expectations explain two further phenomena related to functions. First, they explain what we are asserting when we attribute malfunctions: We are asserting merely that the incapacitated token has failed to behave in accordance with our informed expectations. Second, they explain the forward-looking feature of function attributions: We are expressing the sorts of behavior we anticipate from or predict of future tokens of the functional types. In all of these roles, our attributions are constrained by the bottom-up search for mechanisms that instantiate the attributed systemic capacities. Hence the naturalistic credentials of systemic functions.

IV Malfunctions and Statistical Norms

Before closing my discussion of the naturalistic status of systemic functions, I want to point out that the theory of systemic functions is *not* vulnerable to a surprisingly common complaint. The complaint is that the theory appeals in an implausible way to statistical norms in order to account for the apparently normative status of functions. This worry is misplaced in two ways. First, if I am right that there are no malfunctions among natural traits but only unsatisfied expectations concerning tokens of systemic functional types, then the theory of systemic functions is under no obligation to account for malfunctions. It may be required that we account for our inclination to attribute malfunctions, but we are not obligated to account for that which does not exist. Second, even if my revisionist view of malfunctions is mistaken, the appeal to statistical norms is not required by the theory of systemic functions. Systemic functions are capacities that contribute to the exercise of some higher-level systemic capacity. There is no requirement that functional tokens contribute successfully more often than they fail; the only requirement is that their contributions account in some systemic way for the exercise of the relevant higher-level capacity. I close this chapter with a brief consideration of this second point.

The worry about statistical norms is pressed most forcefully by Millikan (1989) and Neander (1991). Millikan points to examples intended to show that statistical normalcy is irrelevant to the possession of functions. "It is quite possible," she says, "that the typical token of a mating display fails to attract a mate and that the typical distraction display fails to distract a predator" (Millikan 1989, 295). It is assumed that the function of, say, a token mating display is to attract a mate. This is the function of such displays even if the production of displays fails to attract mates more often than it succeeds. The conclusion is that statistical normalcy is not constitutive of functions. This is supposed to tell against the theory of systemic functions and in favor of the theory of selected functions, since selected functions are effects that were selectively efficacious and selective success does not require a positive success rate. It requires only that the trait's effects are successful often enough. The odds of any human sperm cell fertilizing an ovum are low, but because fertilization happens often enough and because its contribution to reproductive success is so great, sperm cells seem to have obvious selected functions but no systemic functions. Or so Millikan claims.

I do not find this convincing. Before giving my main objection, I pause to consider details of the case. We need not agree with Millikan that the function of mating displays is to attract mates in so simplistic a manner. For, it is reasonable to claim that the function of such displays is not to attract mates but rather to increase the probability of attracting mates.

Compare organisms that produce the display to those that do not; such variation must occur between organisms, or must have occurred at some time, for otherwise selection could not sort between them. We will likely discover that those organisms producing the display tend to attract mates and reproduce more frequently than those that do not. What does this show about the function of such displays? We might take it to show that mating displays have the function that Millikan attributes, namely, the function of attracting mates. But we need not take it that way. It might show merely that mating displays have the function of increasing the probability of attracting mates. This leaves room for widespread failures among token displays. It also entails a claim about relative correlations: The production of mating displays is better correlated with reproductive success than the failure to produce such displays. I see no obvious objection to this sort of appeal to statistical norms.

There is a further response to Millikan's charge. Millikan focuses upon token mating displays but gives no argument for individuating behavioral traits so finely. We can individuate behavior as Millikan does, but we need not. Consider, for example, a case in which the females of some species go into season once a year. During this time they produce a specific pheromone that causes the elevation of certain hormones within sexually mature males, resulting in the otherwise absent displays of dramatic colors, or melodious songs, or offerings of food, or whatever. In this case, how should we count the mating displays of the males? One option is to chunk behavior by season, so that each mature male produces a single display that lasts the entire length of the mating season. After all, if the hormonal rage persists for the entire period and then ceases until the next season begins, that provides some grounds for conceptualizing the male's mating behaviors as part of a single, sustained display. And if we accept this scheme of individuating displays, then the correlation between display and mate attraction is high indeed. Another option is to count the displays of a male as a single display so long as they are directed toward a single proximate female. And, of course, other individuating schemes are available. The point is not to defend any one scheme; that should be done on the basis of details of the organisms involved. The point rather is that we should not blithely accept the fine-grained scheme that Millikan takes for granted.

This is true even in the case of sperm cells. I see no obvious reason to accept that individual cells are the relevant functional units from the point of view of mating or from the point of view of fertilization. Single ejaculates are plausible functional units relative to the capacity to mate. The individual cells that comprise an ejaculate may well have a function at a lower level of analysis, relative to the capacity of the ejaculate as a whole. But that is a different matter. The same point holds with respect to fertilization. The relevant functional unit is the aggregate, not the individual cells. Consider an analogy. When we inhale, alveoli in our lungs absorb oxygen. They typically do not absorb all of the oxygen inhaled; some is exhaled, unused. We thus are not inclined to say of each oxygen molecule that it has a systemic function involved in respiration. We do not individuate that finely. Why think otherwise about sperm cells and fertilization? It is more plausible that the function of sperm relative to fertilization, like the function of oxygen relative to respiration, is properly attributed to the aggregate. It is the aggregate that gets the job done. Of course, getting the job done involves systemic capacities of the various lower-level components. So, perhaps the function of any one sperm cell is to contribute to the mobility of the aggregate, in which case the majority of cells regularly fulfill their function. At any rate, and absent a compelling argument to the contrary, there are no good grounds for thinking that each sperm cell has the function of fertilizing an ovum. We should not suppose that sperm are functional in the way Millikan supposes.

In the end, though, none of these considerations bears on the theory of systemic functions, for statistical norms are quite irrelevant to the attribution of systemic functions. What is required for possession of systemic functions is success in contributing to higher-level systemic capacities; but there is no requirement that the incidence of success outnumber the incidence of failure. To see this, consider the relevant system and systemic capacity in the case of mating displays. I take it that the systems involved are potential pairs of opposite-sexed organisms. I also take it that the higher-level systemic capacity we wish to explain is the act of mating or perhaps the production of offspring. A systemic functional analysis of this system will identify traits of the male and female involved in the exercise of this higher-level capacity, including behavioral traits involved in securing a mate. For concreteness, let us suppose that one such behavior is the singing of a distinctive song. Our question, then, is this: In order for the singing of this song to qualify for a systemic function within this kind of system, is it necessary that token singings succeed in attracting mates more often than they fail? The answer is No. All that is required is that, in the context of the specified system, the singing of this song contributes to the exercise of the specified systemic capacity. And a type of behavior can contribute to a higher-level capacity even though its tokens fail more often than they succeed. It may take a dozen recitals to finally arouse the interest of a female, but arousing her attention is precisely the systemic contribution of such singing relative to the larger systemic capacity of mating. The singing of this song will appear in our systemic functional analysis of the system; it will appear in our taxonomy of functions because it contributes in a significant way to the act of mating and the production of offspring. Systemic efficacy is the key to acquiring systemic functions. Such efficacy does not require the statistical norms to which Millikan objects.

Neander (1991) argues that statistical norms are irrelevant to functions for the simple reason that, if they were required, then an epidemic that destroyed all or most token capacities would result in the obliteration of the functional property. If a virus infected our eyeballs and rendered all of us blind, we would be forced to conclude that eyes are no longer for seeing. And that, according to Neander, is nonsense. Eyes would be for seeing even if every member of the population were so afflicted. An epidemic would result in widespread malfunction, to be sure, but that is to say that the functional standing of our eyes would not be lost. Neander concludes that an ahistorical theory of functions, like the one endorsed here, cannot be correct.

Now, it must be conceded that Neander's point has obvious intuitive pull. That, however, is hardly decisive. To see this, notice first that much depends on the details of the story. Suppose, for example, that the virus attacks and devours the neurological connections involved in vision without devouring the eyeballs themselves. In that case, the larger system within which eyeballs once functioned is now destroyed. Our eyeballs would be "systemic components" in name only, without a system in which to operate. Like Aristotle's severed hand, our eyeballs would have lost their functional status because they no longer contribute to the exercise of some higher-level capacity. Suppose further that the disease also attacks the genotype that codes for the visual system, with the effect that our offspring develop "eyeballs" but fail to develop the neurological apparatus required for vision. In this case, phenotypic and genotypic components of the visual system have been destroyed and henceforward human eyeballs would have no systemic function. And yet, on Neander's version of selected functions, these "eyeballs" would retain their functional status, at least for a few generations. For selected functions are constituted by selective history; current capacity is irrelevant. The eyeballs of our offspring, despite the absence of any visual system, nevertheless *would be* descended from selectively successful ancestral eyeballs. So, advocates of selected functions are stuck attributing functions to eyeballs in organisms that lack crucial pieces of the genetic and neurological equipment involved in vision.[9] This is akin to insisting that Aristotle's severed hand would retain the function of enabling the organism to grasp and manipulate objects, even after the complete annihilation of all human appendages. No doubt some of us, upon being afflicted with this dreadful disease of the eyes, would point toward our faces and lament, "These things are *supposed to* enable us to see!" But that, at best, expresses an unfulfilled expectation concerning the status of our abandoned "eyeballs."

Alternatively, suppose the disease does not destroy the larger system of neurological connections nor the relevant genotype. Suppose it merely interferes in some way with the capacity of our eyeballs to process light.

9. Neander can appeal to the proposal of Griffiths (1993) that purports to show that vestigial traits (like the human appendix) do not retain their functional status. Griffiths proposes that functional status is lost so long as enough time has passed to allow for the cleansing effects of regressive evolution. (See chapter 7 for fuller discussion of this suggestion.) But this will not save Neander's view from embarrassment. By hypothesis, the genotypes that code for the larger visual system have been destroyed by the virus, while the genotypes that code for eyeballs remain intact. Now, regressive evolution will lead to the loss or atrophy of eyeballs only if there is variation among the alleles that code for eyeballs. Such evolution will require two or more generations. So even if we endorse Griffiths's proposal, we nevertheless must grant that our eyes (those afflicted with the virus) and those of our immediate offspring continue to possess the selected function of enabling us to see, despite the fact that the larger visual system has been destroyed. This, I claim, puts pressure enough on the theory of selected functions.

In that case, the diseased tokens have lost their capacity to contribute to the larger systemic capacity of seeing. On the theory of systemic functions, therefore, these tokens are rendered nonfunctional. They do not malfunction for they no longer belong to the relevant functional type. They are now devoid of their former functional status. This is true even if the disabling effects of the disease are temporary. So long as the token lacks the capacity to fulfill the specified task, it is not a member of the functional type.

Here, however, is where Neander's view exerts its force, for we may feel that our diseased eyes belong to their former functional type despite the destruction of the underlying capacity. We may feel inclined to categorize these tokens as merely malfunctional as opposed to nonfunctional. This is even more intuitive if we suppose that the affliction is temporary, for then it appears that I am forced to the unintuitive view that the relevant tokens had a function, then lost it, then regained it.

Again, I concede the prima facie power of the intuition. Our folk biological or physiological intuitions incline us to think that diseased eyes are malfunctional and not merely nonfunctional. But precisely here is where I believe that our concept of functions must be revised. As I argued in chapter 5, the explanation of malfunctions offered by advocates of selected functions commits them to a dilemma: Either they fail to account for malfunctions or they endorse an ontology at odds with our naturalistic orientation. This should make us question the intuitions that seem to favor Neander's position. I also suggest that certain of our psychological capacities incline us to see some natural traits as "properly" functional or genuinely "teleological" and others as merely useful. In particular, our inclination to see some tokens as malfunctional can be explained by appeal to our informed expectations of systemic functional categories. This Humean strategy provides speculative grounds on which to explain away the intuitions to which Neander and others appeal.

V Conclusion

But there is another reason for rejecting Neander's intuitions: The theory of selected functions, despite its claims to the contrary, cannot account for the possibility of malfunctions. The theory endorsed by Neander and others cannot substantiate the central intuitions to which Neander and others appeal. This is a point about the internal resources of the theory of selected functions—there are no grounds within the theory to warrant the attribution of malfunctions. I defend this claim in the next chapter. The general lesson is that the intuitions on which Neander and others rely—intuitions concerning the norms of nature—have no standing. We ought to reject the lingering sense we have that natural traits have functions that are "proper" or genuinely "teleological." We do better to change the way we understand the nature of functions: Nonengineered, natural traits do not and cannot malfunction. As naturalists, we should not pretend otherwise.

7

Defining Functional Types: Success and Malfunctions

Of all the claims put forth by advocates of selected functions, perhaps the grandest is the claim that their theory provides a clear and compelling naturalistic explanation for the occurrence of malfunctions. Selected functions, it is claimed, are inherently normative; selected functions contain a built-in norm of evaluation that makes possible the occurrence of selected malfunctions. The same claim is made on behalf of historical functions generally, including the theory of weak etiological functions proposed by Buller (1998). The basic thought is that some sort of success among ancestral tokens imposes a standard or norm of performance upon descendent tokens of any trait. Success in selection is typical, but success in contributing to fitness is thought to suffice as well. Either way, the result is that descendent tokens are now "supposed to" perform the selectively successful tasks, so that when a token fails to so perform—when damage or disease destroys the requisite physical capacities—that token nevertheless retains its functional role. Precisely this, it is claimed, makes possible the occurrence of malfunctions.

Advocates of the historical approach further claim that accounting for malfunctions is one respect in which their theory is clearly superior to the theory of systemic functions. Systemic functions, it is alleged, can be attributed to items that intuitively have no biological purpose, that do not enjoy the standards or norms of performance that historical success confers. Millikan's claim concerning the function of rain clouds, discussed in chapter 4, illustrates this charge. And insofar as systemic functions are nonpurposive, they cannot capture the normative force of functional properties. The poverty of systemic functions contrasts starkly with the riches of selected functions. This alleged inferiority of systemic functions has led some theorists to reject the theory outright and prompted others to opt for some version or other of the combination approach to functions. The current consensus, at any rate, is that any defensible account of functions must include the theory of selected functions, or appeal to some sort of historical success, in order to account for the norms of nature.

Some advocates of the historical approach claim still further that the theory of selected functions is fruitfully applied to various philosophical problems in other areas. Historical functions, it is claimed, insofar as they make possible the occurrence of malfunctions, provide standards of correctness and incorrectness of various sorts. Millikan (1984, 1993), Papineau (1984, 1987, 1993), Neander (1995) and others apply the theory of selected functions in philosophy of mind. They claim that the normative status of selected functions is key to understanding the occurrence of misrepresentations. Malfunctions among cognitive mechanisms give rise to mistaken representations in a variety of attitudes—beliefs that are false, desires for what is unreal or impossible, and so on. Selected functions are similarly applied to the mechanisms involved in various normative judgments.[1] In all these cases, the purported capacity of selected functions to persist in the face of physical incapacitation is crucial to accounting for the relevant phenomena. The theory of selected functions, then, appears to bear fruit where the theory of systemic functions cannot.

The thesis of this chapter is that none of these claims is warranted. I shall argue that the historical approach to functions lacks the resources with which to account for the possibility of malfunctions. Historical malfunctions are impossible. Both theories—selected functions and weak etiological functions—fail to account for the alleged normative status of functions. And this, I argue, is an internal failing of the historical approach: Both the theory of selected functions and the theory of weak etiological functions lack the internal resources with which to account for the possibility of malfunctions. The consequences are important. Failure to account for malfunctions eliminates a central motivation for adopt-

1. Additional appeals to the alleged normativity of selected functions in philosophy of mind include Dretske (1995), McGinn (1989), Post (1991), and Rowland (1997). Appeals in epistemology include Lycan (1988).

ing the combination approach. It also defeats the claim that historical functions give rise to standards of correctness or incorrectness. Thus, the application of selected functions in philosophy of mind, epistemology, and ethics cannot get off the ground.

A further consequence is of especial importance. In chapter 3, I claimed that the theory of systemic functions, applied to populations, underwrites all the function attributions warranted by the theory of selected functions. The theory of systemic functions can explain everything that the theory of selected functions can explain—and more besides. Or so I claim. One way to resist this claim is to insist that there is at least one thing explicable from within the theory of selected functions (or the theory of etiological functions) alone, namely, the possibility of malfunctions. If the theory of selected functions, but not the theory of systemic functions, justifies the attribution of malfunctions, then we have strong prima facie reason for retaining the former as a separate, autonomous theory. The burden of this chapter, then, is to show that this line of resistance is illusory. Neither version of the historical approach contains the internal resources with which to warrant the attribution of malfunctions. Selected malfunctions and weak etiological malfunctions are impossible. Or so I shall argue. And if my argument is sound, we have no reason whatsoever to preserve the historical approach to functions.[2] The theory of systemic functions is all we need.

A further thesis is that a parallel conclusion concerning malfunctions holds for the theory of systemic functions—though without the embarrassing fallout. The theory of systemic functions does not define functional kinds in terms of historical success. Systemic functional types are defined, rather, by reference to capacities that contribute to some sort of systemic success. In consequence, the loss of a systemic capacity entails loss of the corresponding systemic function. This is not a criticism, however, because the theory does not pretend that functional properties are normative. Moreover, the theory of systemic functions can explain why we are inclined to see natural traits in terms of such norms. The

2. This is not to say, however, that the theory of selected functions should be embraced if the argument of this chapter is unsound. Criticisms offered in previous chapters are enough to motivate the search for an alternative to the historical approach.

speculation is that we mistakenly attribute norms of performance to trait tokens, when in fact the only norms involved are epistemic in character, underwritten by our expectations concerning functional types and generated by the functional categories we discover in the course of applying the theory of systemic functions.

I Selected Malfunctions

I begin my argument with selected malfunctions and extend it to weak etiological malfunctions thereafter. The theory of selected functions, recall, asserts that the selected function of T in organism O in selective environment E is to do task F if and only if:

(i) Ancestral tokens of T in O performed F in E,
(ii) T was heritable,
(iii) Ancestral performances of F enabled organisms with T to satisfy demands within E better than organisms without T,
(iv) Superior satisfaction of selective demands enabled organisms with T to out-reproduce (in the long run) those without T,
(v) Superior reproduction caused organisms with T and hence tokens of T to persist or proliferate in the population,
(vi) Superior reproduction among organisms with T, by causing the persistence or proliferation of tokens of T, imposes an office or role upon descendents of T.

Condition (v) asserts the existence of a causal-historical relation between ancestral traits selected for and descendents of those successful ancestors. This relation is typically characterized in explanatory terms. The claim is that the selective efficacy of traits described in (i)–(iv) constitutes the explanans and the persistence or proliferation of traits mentioned in (v) constitutes the explanandum. The attribution of a selected function, on this construal, is equivalent to the selective explanation of the persistence or proliferation of the specified trait. As indicated in chapter 2, however, this explanatory relation is parasitic on the relevant ontological relations. Conditions (i)–(v) are explanatory only because the selective efficacy of the effects cited in (i)–(iv) caused, via the mechanisms of selection, the persistence or proliferation of traits cited in (v). It is because ancestral instances were selectively efficacious that latter-day instances persisted or proliferated, and this causal relation warrants the explanatory relation between (i)–(iv) and (v).

The theory of selected functions also asserts that latter-day tokens of trait T possess the relevant selected function even when, due to incapacitation, they are causally unable to fulfill that function. Thus, condition (vi). The justification for this assertion, discussed in chapter 2, is persuasive on its face: The causal relation between conditions (i)–(iv) and condition (v) is entirely historical. All that matters for acquisition of a selected function is that present tokens are descended from selectively successful ancestors. In particular, current capacities are irrelevant. Whether or not a descendent token can perform F is quite beside the point so long as it is related by descent to a selectively successful lineage. After all, present tokens of T do exist, at least in part, as a result of the selective efficacy of their ancestors. This is true no matter what the capacities of these tokens.[3] Hence the claim in condition (vi) that present tokens are endowed with the "role" of fulfilling the selectively successful tasks and, in consequence, that these tokens are "supposed to" do whatever their ancestors did that resulted in selective success.

There is, however, an apparent tension within the theory of selected functions between malfunctioning traits on the one hand and vestigial traits on the other. This tension has not gone unnoticed. The above characterization of selected malfunctions suggests that a token trait has a selected function just in case it is descended from a selectively successful lineage. But this is too liberal. If descent from a successful lineage were both necessary and sufficient for possession of a selected function, we would be forced to accept that vestigial traits such as the human appendix retain their selected functional status. This consequence is unintuitive and has been disavowed by at least one advocate of selected functions. Griffiths (1992) defines selected functions with the express intention of ruling out vestigial traits and ruling in malfunctioning traits. While his basic formulation of the theory is equivalent to conditions (i)–(v) above, he adds the further requirement that these conditions must apply to trait T during the most recent "evolutionarily significant period":

3. At least some advocates are explicit on this point. See the passage from Millikan (1989) quoted in chapter 5, section IV.

An evolutionary *[sic]* significant time period for trait T is a period such that, given the mutation rate at the loci controlling T, and the population size, we would expect sufficient variants for T to have occurred to allow significant regressive evolution if the trait was making no contribution to fitness. A trait is a vestige relative to some past function F if it has not contributed to fitness by performing F for an evolutionarily significant period. (Griffiths 1992, 128)

The suggestion is that if recent ancestors of T have been selectively efficacious, then current tokens of T have not been subject to regressive evolution, in which case they retain their functional status. But if recent ancestors have not been selectively efficacious, and if the loci that code for T produced variants of T, then it is probable that recent ancestors have been subject to regressive evolution, in which case they qualify as vestiges. Hence the requirement that (i)–(v) apply to ancestors of T in the most recent period during which evolution could have occurred—to rule out the possibility of regressive evolution and ensure the retention of functional status. Thus, while (i)–(v) may apply to the effects of the human appendix from a distant period of selection and evolution, they do not apply in any recent periods and, in consequence, present-day tokens do not possess the earlier selected function.[4]

The virtue of this proposal is that, while apparently ruling out vestigial traits, it leaves ample room for malfunctioning traits. We need only consider lineages of traits selected for during the most recent evolutionarily significant period. Tokens descended from a recently or presently successful lineage may malfunction as a result of various causes. Perhaps the mammalian heart is like this. Suppose the heart contributed to relative fitness during recent periods of selection. Nevertheless, a token heart may lose the capacity to fulfill its functional task. A congenitally diseased heart

4. Does this proposal save advocates of selected functions from the problem of vestigial traits? Griffiths's aim is to define functional categories that include malfunctioning tokens and excludes vestiges. But suppose that ancestors of T performed F in a selectively efficacious manner during the most recent evolutionarily significant period. Suppose further that, just as the recent period of evolution came to a close, the organism's internal economy was altered in some small way so that T no longer does F or that T's doing F is idle. This may occur when a mutation affecting some phenotype other than T renders T's performance of F superfluous. Regressive evolution has not had time to occur and yet it seems plausible that current tokens of T are vestiges. They are vestiges because they no longer contribute to the organism's economy. This is true despite the fact that we must cite the selective efficacy of recent ancestors to explain the existence of current tokens.

may fail to develop into a muscle with the requisite capacity to circulate blood in the normal way, while a damaged heart may lose the requisite capacity. Of course, some forms of incapacitation are transitory. A diseased heart, for example, may recover enough muscular capacity to circulate blood in the normal way thanks to surgical or therapeutic intervention. Nonetheless, whatever the particular cause and whatever its duration, a nonvestigial trait may lose the requisite physical capacity and thus qualify as malfunctional. Or so it is claimed.[5]

On Griffiths's suggestion, then, we may revise the theory so that possession of a selected function requires genealogical descent from a recently successful lineage. Following his formulation, we may assert that

> (SF) A token of trait T has selected function F if and only if the token is descended from a lineage perpetuated by the recent selective efficacy with which ancestral members performed F and now occupies the appropriate selectively generated office or role.

Now, I want to leave open the question whether or not other qualifications are in order here. Perhaps, upon reflection, we will discover that descent from a recently successful lineage is not sufficient and other requirements ought to be included. What matters for present purposes, however, is that the conditions under which a token trait has a selected function definitely do not include possession of the capacity required to actually perform the functional task. The requisite physical capacity may be absent but the functional property persists; precisely this makes

5. It should be clear that I am concerned with cases in which disease or damage renders a token trait unable to perform a specific functional task. There may be cases in which partial damage diminishes a trait's performance but does not incapacitate it altogether. We might be tempted to say that the trait is partially malfunctional. Now, some cases may strike us this way only because we do not know the exact function involved. The intuitive function of the heart is to pump blood, but it is likely that the function is to pump blood at a certain rate and with certain force depending on a variety of conditions internal and external to the organism. By contrast, there may be instances in which the attribution of a malfunction is genuinely vague. The focus of this discussion, however, is cases in which incapacitation unambiguously inhibits the performance of the functional task. My aim is to demonstrate that the historical approach fails to account for the attribution of malfunctions where the failure in performance is clear.

possible malfunctioning traits. And this, precisely, is what I wish to challenge.

II Selected Functional Types

Griffiths's revision helps, but there remains a problem with the theory of selected functions. Conditions (v) and (vi) refer to "tokens of trait T" and the extension of this phrase is ambiguous. Advocates of selected functions construe its extension broadly while I construe it narrowly. After defending my narrow construal, I shall show why selected malfunctions are impossible.

Any occurrence of evolution by natural selection requires at least one trait that is the target of selection and at least two variants of the targeted trait, one selectively successful, the other relatively unsuccessful. In the case of the peppered moth (see chapter 2, section I), wing coloration was the targeted trait, for there was selective discrimination among varying shades. Dark and light coloration were the relevant variants and only one was selected for. I shall refer to the targeted trait type as the "generic trait type" and the selectively successful variant type as the "selected functional type." To generalize, any occurrence of evolution by natural selection requires that there exists

(a) at least one generic type of organismic trait—the target of selection—containing variation,

(b) at least one narrower type within the generic type comprising variants that are not selected for, and

(c) at least one narrower type within the generic type—the emerging selected functional type—comprising variants selected for.

Since all three are required by the theory of evolution by natural selection and since selected functions are the products of natural selection, all three are required for acquisition of a selected function.

Significantly, (a)–(c) are satisfied when variants of the generic type include the incapacitated—when diseased or damaged instances satisfy (b) and nondiseased or undamaged satisfy (c). Consider a distant ancestral population in which a precursor to the human heart first emerged. Suppose that, prior to the emergence of this proto-pump, nutrients were distributed and wastes collected by way of some sort of diffusion and secretion device. Relative to the generic trait type "circulatory mechanism" the population contained two morphologically distinct variants, proto-pumps and diffusion/secretion devices. Proto-pumps, we may suppose, enjoyed relatively greater selective efficacy by virtue of superior efficiency. Within each variant, however, the population contained further variants also of selective significance. Relative to the type "proto-pump," there were diseased and nondiseased variants, and also damaged and undamaged variants. Within the type "diffusion/secretion device," there likewise was variation between diseased and nondiseased variants, and so on. These latter variants, of course, were selectively significant, since the difference between them tended to result in substantial differences in reproductive output.[6]

Of course, (a)–(c) do not require the variation provided by incapacitation. Other sorts of variation suffice. Conditions (b) and (c) require only that some variants are selected for while others are not. The only requirement is differential output caused by differences in traits and this requirement is met when, for example, one intact trait (proto-pumps) outperforms a second intact trait (diffusion and secretion devices). But, of course, it is a bald biological fact that diseased and damaged tokens are ubiquitous and that such tokens provide one sort of variation upon which selection in fact acts. The de facto importance of incapacitation is enough to make it important in understanding the theory of selected functions.

This distinction between generic trait categories and selected functional categories bears directly on the ambiguity in "tokens of trait T" in condition (v) of the theory. For, generic types and selected functional types are not coextensive. The central difference is that selected functional categories are individuated and defined by reference to causal effects that were selectively successful. Selected functional types, in a word, are success types. Generic types, by contrast, are not. Indeed, generic categories, out

6. The obvious point here is that incapacitated variants tend not to contribute to selective success. What are the causes of incapacitation? At a minimum, disease and damage. But we must take care not to read "diseased" or "damaged" as inherently normative, as smuggling norms of performance in through the side door. They should be read, rather, in terms of the informed expectations we bring to bear on tokens devoid of systemic capacities. See chapter 6, section III, for discussion. We must also bear in the mind that an incapacitated token of T is no longer a member of the category of Ts, *if* the category is defined by reference to some systemic capacity. If the category is defined otherwise—for example, in terms of homologies—the token may retain its membership in that type.

of which functional categories emerge, contain both the damaged and the nondamaged tokens, whereas functional categories contain only those tokens that were sufficiently intact to contribute to selective success. So the ambiguity afflicting condition (v) can be put in the form of a question: Are the tokens of trait T cited in (v) members of the relevant generic type only, or are they also members of the emerging selected functional type? Does the satisfaction of (i)–(iv) warrant the assertion of condition (vi) along with the assertion of (v)?

The importance of this question is considerable. If tokens of T cited in (v) are members of the generic type only, then, since generic types include instances that are incapacitated, selected malfunctions are possible. That is, if condition (v) classifies together the damaged with the nondamaged, the diseased with the nondiseased, and so on, then tokens that can no longer perform the associated functional task nevertheless belong to the functional category, in which case malfunctions are possible. But if tokens of T in (v) are members not only of the generic type but, more narrowly, members of the selected functional type, then selected malfunctions are impossible. Selected functional types are individuated in terms of the property selected for—in terms of selective success—and thus any token that is so diseased or so damaged as to lack the relevant property selected for is thereby excluded from the selected functional category. In that case, however, selected malfunctions are impossible, since exclusion from the selected functional type precludes the possibility of selected malfunction. What, then, is the correct answer?

The answer is that tokens of T cited in (v) are not merely members of the generic category but, more specifically, members of the narrowly circumscribed selected functional category. To see why, suppose that this were not so. Suppose "tokens of T" in (v) referred to the broader generic type. In that case, we would be guilty of equivocating between conditions (i)–(iv) on the one hand and condition (v) on the other. For, "tokens of T," as employed in (i)–(iv), refers unequivocally to properties in terms of which the emerging functional category is defined. It refers solely to properties selected for and hence solely to members of the incipient selected functional type. On pain of equivocation, therefore, "tokens of T" in condition (v) also must refer to members of that selected functional type. And this means that membership in a selected functional category, per condition (v), requires possession of the property selected for.

It is perhaps tempting to object that the above argument is persuasive only if we take seriously the worry about equivocation, but that we need not take it seriously. After all, the selective success of past tokens of T caused the persistence of the entire organism and hence of the all the organism's parts. So the selective efficacy of traits cited in (i)–(iv) explains the persistence of much more than just tokens of T endowed with the property selected for. It explains, in particular, the persistence of the disposition for certain sorts of disease and damage. Indeed, to the extent that past hearts were selectively successful, they contributed to the perpetuation of the entire genotype for hearts and no doubt this genotype brings with it dispositions toward various forms of incapacitation. If so, then we can indeed explain the persistence of diseased and damaged hearts in the same way we explain the persistence of healthy hearts. Worries about equivocation thus are misplaced because they artificially restrict the causal or causal-explanatory force of the theory of selected functions.

I agree that we can explain the persistence of tokens subject or prone to incapacitation in evolutionary terms. I do not agree, however, that this is relevant to the theory of selected functions. The reason is twofold. First, I agree that the selective success of past instances of T caused the persistence of the entire organism and hence the organism's parts, including dispositions toward disease and (some forms of) damage. But the persisting parts also include kidneys, lungs, the liver, the brain, and much more. The causal-historical relation between successful ancestral hearts and the persistence of the entire organism includes a causal-historical relation between successful ancestral hearts and latter-day kidneys, lungs, livers, brains, etc. But the mere existence of this relation does not show that kidneys, lungs, livers, etc. belong to the selected functional kind defined in terms of circulating blood; the mere existence of this relation does not endow these other traits with the selected function belonging to the heart. Hence the causal-historical and causal-explanatory relations between ancestral hearts and present-day traits does not settle the issue, in which case the worry about equivocation is a legitimate one. The theory of selected functions must restrict the extension of "tokens of trait T" as it occurs in condition (v) in a principled way grounded in the theory of evolution by natural selection.

But, of course, the theory of selected functions does individuate functional traits in a principled way—by appeal to selective success, by appeal to properties selected for. This brings me to my second reason for rejecting the above objection. Conditions (i)–(v) of the theory identify selected functional properties from among organismic properties generally: Selected functions are properties that have persisted or proliferated as a consequence of their own selective efficacy; they are properties that perpetuated themselves insofar as they were selected for. Crucially, it is by reference to these *self-perpetuating effects* that selected functional categories are individuated in conditions (i)–(iv). Therefore, insofar as tokens of T cited in condition (v) are truly said to have a selected function—insofar as they belong to the functional type described in (i)–(iv)—they must possess the qualifying property of that particular functional category. So, selected functional traits must possess the relevant property selected for—in which case the extension of "tokens of trait T" as it occurs in (v) is narrow, excluding those tokens that lack the relevant property selected for.

As we have seen, the possibility of selected malfunctions hangs on the claim that selected functions persist even when the physical capacity is lost. This, in turn, hangs on the further claim that selected functions, being entirely historical in nature, in no way depend upon current capacities. This claim is expressed in (SF) above. The lesson of this section, however, is that selected functional types are individuated in terms of properties that, by virtue of being selected for, perpetuate themselves across generations. Hence, membership in a selected functional type requires more than descent from a recently successful lineage; it also requires possession of the individuating property. So, rather than (SF), we require

> (SF*) A token of trait T has selected function F if and only if (1) the token is descended from a lineage perpetuated by the recent selective efficacy with which ancestral members performed F and (2) the token possesses the property selected for in terms of which the relevant functional category is defined and, in consequence, now occupies the appropriate selectively generated office or role.

And if, due to some sort of disease or damage, the token of trait T lacks the property selected for, then it is merely a member of the generic type, not the functional type. This is why, as I shall now argue, selected malfunctions are impossible.

III Why Selected Malfunctions Are Impossible

The argument for the impossibility of selected malfunctions rests upon three claims. First, a token of trait T, as a consequence of congenital disease, can fail to acquire a property selected for. Or a token can lose a property selected for as a consequence of damage. It is possible, in short, to lose or to fail to acquire the property in terms of which the associated functional type is defined. Suppose, to be blunt, that we inject the wings of a dark-colored moth with a solvent. Dark coloration is lost. So too, then, is the property selected for.

Second, loss of the property selected for disqualifies the token from the associated selected functional category. For such types are individuated in terms of a property selected for; loss of the defining property entails loss of membership in that category, as the second condition of (SF*) makes explicit. So, destruction of dark coloration in the moth destroys its capacity for camouflage and thereby destroys its possession of the property selected for. Of course, the loss of dark coloration does not make its wings colorless and hence the color of the wings still belongs to the generic trait type "being colored." But that is not sufficient for membership in the selected functional type "providing camouflage."

The claim that incapacitated instances no longer qualify as members of the relevant selected functional category applies quite generally. The category "mammalian heart," for example, constitutes a generic type, not a selected functional type. Some ancestral hearts were incapacitated and thus unable to pump properly, while others were intact. The former were selected against while the latter were selected for; more precisely, the former lacked the property selected for while the latter enjoyed it. So the category "heart," which contains the efficacious and the impotent, cannot be a selected functional type. An incapacitated token today, therefore, despite its descent from a recently successful lineage of hearts, lacks the defining capacity and hence cannot be a member of the relevant selected functional type.

Advocates of selected functions have missed this point. In particular, Millikan (1984, 1989) and Neander (1991, 1995) claim that several generic categories of traits, including the type "heart," are essentially selected functional categories individuated according to properties selected for. But that is false. In fact, it is impossible given the biological reality of diseased and damaged hearts. For even when hearts first emerged on the evolutionary scene, the nondiseased and undamaged were selected for and the others selected against. The relevant selected functional category, then, is defined by reference to the effects of those that were nondiseased and undamaged. The generic category, which includes instances lacking the property selected for, should not be mistaken for the functional kind.

This is not to say that there is no selective explanation for the existence of generic types. Such explanations exist. But it is an error to assume that all selective explanations entail the attribution of selected functions. Traits selected for the efficacy of their own effects qualify for selected functions, but traits that persist or evolve merely by virtue of their connection with some other trait do not qualify for selected functions. Such traits are merely selected of. There is a selective explanation for such traits—one that appeals to the selective efficacy of other properties of the organism—but not an explanation that warrants the attribution of functions to traits selected of. This is how it is with generic types. Generic types, by virtue of their connection to selected functional types, are selected of. They are the happy by-products of the self-perpetuating traits with which they are connected.

This leaves open the question, How are generic types individuated? If generic types are not selected functional types, then what are they? I am prepared to leave this question unanswered, since the difficulty to which it points is a challenge to advocates—not critics—of selected functions. After all, the theory of selected functions asserts that functional traits are a product of their own selective efficacy. But, as we have seen, evolution by natural selection occurs only when there is variation within the generic category and only when there is selection for certain variants within that category. So, advocates of selected functions, in defense of their theory, owe us an adequate characterization of the individuation of generic types.

Moreover, Amundson and Lauder, in their splendid (1994) paper, address the issue of individuating traits on nonfunctional grounds. They point out that many significant categories of traits are classified not in terms of function but rather in terms of homology. Distinct traits are homologous so long as both descend from the same feature of a common ancestor. And we recognize two traits as instances of a single homologous category by virtue of their structural properties. Thus, for example,

> the wing of a sparrow is homologous to the wing of an owl because the character "wing" (recognized by a particular structural configuration of bones, muscles, and feathers) characterizes a natural evolutionary clade (birds) to which sparrows and owls belong.

Significantly, traits can be homologous and nevertheless differ in their function:

> The forelimbs of humans, dogs, bats, moles, and whales, and each of their component parts—humerus, carpals, phalanges—are homologous. Morphologically they are the same feature under different forms. Functionally they are quite distinct. (Amundson and Lauder 1994, 454)

Perhaps, then, homologies are one sort of generic type upon which selection acts and out of which emerge various functional types. No doubt the suggestions offered by Amundson and Lauder are not intended as a full-blown account of generic trait categories. Nevertheless, they provide an informed strategy that should allay the worry that my talk of generic types is biologically naive or ungrounded. Generic types can be individuated nonfunctionally in terms of homologies and, however individuated, generic types are necessary for evolution via selection. And, of course, categories individuated in terms of homology are distinct from categories individuated in terms of selective success, since homologies include members that are damaged and diseased.[7]

The third and final premise of my argument is that incapacitated traits, insofar as they do not qualify as members of the selected functional type, cannot possess the corresponding selected malfunction. I assume that no trait can possess a selected malfunction unless it possesses the corresponding selected function. So incapacitated tokens of trait T cannot possess selected malfunctions; they are nonfunctional, not malfunctional. I conclude, therefore, that selected malfunctions are impossible.

It might be thought that this conclusion is too strong. For it might be thought that one way a trait can possess a selected malfunction is by

7. It is perhaps worth emphasizing that selected functional kinds are narrower than homologies. The functional category is defined by reference to ancestral tokens that were selectively successful while homologous categories are defined entirely in terms of genealogy and not in terms of any sort of success.

environmental dislocation. Being removed from conditions of normal functioning may result in malfunction despite the absence of disease or damage. Since my argument at best shows the impossibility of selected malfunctions due to incapacitation, then my conclusion must be tempered to leave open the possibility of malfunction due to dislocation. Now, this objection assumes that the failure to fulfill a function as a result of dislocation results in malfunction. But I do not think this is true. Traits can cease functioning without malfunctioning. If we remove a healthy heart from a fresh corpse and rush it to the hospital for transplant, there is a period during which it is environmentally dislocated and does not pump blood. But it is not malfunctioning. The heart's failure to pump, in this scenario, is not due to any change within the heart. It is due solely to dislocation from its normal environment. Just as it is implausible that, say, an intact electrical appliance removed from any source of electricity is thereby malfunctioning, so too it is implausible that an intact heart removed from its normal conditions of functioning is thereby malfunctioning.

But suppose I am mistaken; suppose dislocation can result in malfunction. Still, the force of this objection is weak. Any theory of functions that can explain malfunction due to dislocation but not due to incapacitation is an overly narrow theory of functions. It would fail to account for paradigm cases of malfunctions—discoloration in wings, hearts with damaged valves, and the like. And the narrowness of the theory would infect the standards of correctness it aspires to provide in epistemology, ethics, philosophy of language, and philosophy of mind. So this objection, I conclude, does not stand. Selected malfunctions are impossible.

IV Systemic Functions and "What for?" Questions

It might be objected that not all versions of the theory of selected functions fall prey to the above argument. Advocates of selected functions tend not to emphasize the fact that selected functional kinds are essentially success kinds. Moreover, some advocates never assert that the theory can account for the occurrence of malfunctions or that selected functions are normative in any explicit sense. It thus might be thought that the above argument, if sound, is insufficiently general to force upon us the rejection of the theory of selected functions.

There is little to say in response to the first point. It is true that most advocates of selected functions do not explicitly define their functional kinds as I have suggested. But part of the burden of the last two sections is to show that any version of the theory is indeed committed to the claim that selected functional kinds are defined in terms of selective success. Whether or not advocates of the theory have been explicit in this regard is irrelevant. The second point is more interesting. Brandon (1981, 1990), for example, is silent on the issue of malfunctions or functional norms, and this might give the impression that there is a version of the theory immune to the argument of this chapter. I want to consider this possibility briefly.

Brandon's approach to the issue of functions is unique. Drawing on the erotetic model of scientific explanation, he suggests that scientific theories are individuated in part by the questions they answer. Most theories are constructed to answer a "why?" question of some sort and can be distinguished, at least to some extent, on this basis. Crucially, in addition to asking "why?" questions, evolutionary biologists also ask "what for?" questions:

> we might wonder about the song of humpback whales, the color patterns of penguins, the lack of leaves in cacti, or the nastiness of rats in crowded conditions. We might verbalize our questions using "Why," "How come" or "What for." I'll refer to such questions as *what-for* questions. . . . *What-for* questions are, apparently, teleological. (Brandon 1981, reprinted in Brandon 1996, 33)

Now, Brandon does not put much stock in the specific terms with which we cast our question; he explicitly is not engaged in any form of linguistic analysis. But he does suggest that the question "what for?" often enough indicates a request for an explanation that is teleological in character. And he goes on to assert that "[t]oday biologists, and most philosophers of biology, agree that Darwinian answers to what-for questions are legitimate" (33). Brandon's view, then, is that we can answer the question, What is this type of trait for? by appeal to the theory of evolution by natural selection. The answer, in particular, has the form described in the theory of selected functions. A trait is for whatever it was selected for; a trait is for whatever task was selectively efficacious among ancestral tokens of heritable traits. Natural traits are teleological by virtue of descent from selectively successful ancestors.

Although Brandon makes no mention of malfunctions or functional norms, I believe that the argument of this chapter also applies to his version of the theory. For, it is hard to understand the alleged teleological nature of selected functions without attributing norms of evaluation. To say of some trait that it is *for* the performance of some task is, I take it, to say that the trait is supposed to perform the task and thus that there is some standard of assessment that applies to tokens of the type. This means, incidentally, that Brandon is faced with questions about the ontology of selected functions discussed in chapter 5. More to the present point, he is also faced with the argument of this chapter, for the norms of evaluation to which Brandon appeals are explicated in terms of ancestral selective success. He does not explicitly define selected functional kinds as I suggest above, but it is clear from his discussion that functional types are individuated in terms of selective success. And this means that, on Brandon's view, membership in the selected functional kind is lost when the token is sufficiently incapacitated—in which case selected malfunctions are impossible. I conclude that we should reject Brandon's account of selected functions along with the rest.[8]

Interestingly and more generally, the phenomena to which Brandon appeals—the prevalence of "what for?" questions among evolutionary biologists—are entirely explicable from within the theory of systemic functions. The sorts of theorizing to which Brandon appeals in support of selected functions are best understood, I believe, in terms of systemic functions. We simply have no need of the claim that natural traits are teleological or "proper" in order to understand the theoretical importance of "what for?" questions in biology. For, when we (or evolutionary biologists) ask what the color pattern of penguins is for, our question is ambiguous. We might be asking for the "proper" standard or norm that applies to token color patterns. And the correct answer to that question is that color patterns are not for anything. As naturalists, we should eschew the attribution of norms of performance to natural, nonengineered

8. As noted earlier (chapter 5), Brandon (forthcoming) endorses the combination approach and thus rejects his (1990) endorsement of the historical approach. But to the extent that the combination approach incorporates the theory of selected functions, the above assessment applies to Brandon (forthcoming) as well.

traits. But we might be asking a different question. We might be asking for an account of the capacity of these color patterns to contribute to some higher-level systemic capacity. There is no presumption that these patterns are "for" anything, only that they contribute to the exercise of some higher-level capacity. And the correct answer will depend upon details of the system involved. We might inquire into the possible role of such patterns in mate selection. Or we might wonder about the possible role of such patterns in the evolution of the population. The first inquiry would take the subpopulation of sexually mature penguins as the basic system and consider whether color patterns play a role in the formation of mating pairs. The second would take the whole population at some point in its history as the relevant system and consider whether color patterns were among the traits targeted by various selective demands. Other systems and systemic capacities are possible. In all such cases, however, the biologists' "what for?" should be read naturalistically as asking for the salient systemic function of the specified trait rather than some vague sort of teleology.

V Weak Etiological Functions

As discussed in chapter 2, Buller (1998) advocates an alternative version of historically derived functions and malfunctions. He suggests that we dispense with the appeal to selective efficacy and focus simply upon causal contributions to fitness. The core idea is persuasive: Trait T has function F so long as ancestral tokens contributed to some component of fitness—for example, fecundity or viability—and thereby contributed to the perpetuation of the type. This can occur when, for one reason or another, trait T fails to manifest the sorts of variation upon which selection can act (see chapter 2, note 13). The point, then, is that there is a kind of historical success—success in contributing to fitness—that helps explain the persistence of a trait even when the trait has never been the target of selection. The only other requirement is that T was heritable among ancestral organisms. We may formulate the theory of weak etiological functions this way:

(WEF) A token of trait T has weak etiological function F if and only if (1) ancestors of this token contributed to some component of organismic fitness by performing F and (2) T was heritable. (See [ET1] and [ET2] in Buller 1998, 511.)

Appeal to these two conditions helps explain the persistence of the type and does so without recourse to the selective efficacy of ancestral instances.

Eschewing appeal to selection may give the appearance that my argument against selected malfunctions does not apply to the theory of weak etiological functions. But that is not so. The argument does not apply in the exact same way, to be sure, since it is cast in terms of selective success. But the argument cast more generally—in terms of historical success—does apply to the theory of weak etiological functions. On Buller's view, selection is not necessary; contributions to ancestral fitness suffice for weak etiological functions. But, of course, either there was variation among ancestral tokens of T with respect to their contributions to fitness, or no such variation occurred. Suppose there was variation with respect to fitness. Suppose that some variants successfully contributed to fitness while others, due to incapacitation, did not. The etiological functional category, in this case, is individuated by reference to those variants that, in consequence of doing F, in fact contributed to ancestral fitness. The category is defined by reference to success vis-à-vis fitness. So tokens today that, due to incapacitation do not possess the capacity involved in fulfilling task F, do not qualify as members of the functional type. They do not qualify because they lack the defining success capacity. But now suppose that there was no variation in ancestral instances of T. Suppose, more precisely, that not a single variant was sufficiently incapacitated to prevent it from contributing to fitness. The etiological functional category, in this case, is individuated by reference to those tokens—which, as it happens, includes all of them—that in fact contributed to ancestral fitness by virtue of doing F. So, once again, tokens today that, due to some form of incapacitation do not possess the capacity involved in fulfilling task F, do not qualify as members of the functional type. They lack the defining capacity of success. Therefore, weak etiological malfunctions are impossible for the same general reasons that selected malfunctions are impossible.

Of course, the basic point generalizes further, infecting the combination approach to functions (explicated in chapter 2). The problem with the theories of selected functions and weak etiological functions is that they define functional kinds in terms of some type of historical success. But the same problem afflicts all combination theories as well. Preston (1998), for example, claims that selected functions are required in any adequate theory of functions in order to capture the presumed normativity of functions. She goes on to advocate a view embracing both systemic and selected functions. The lesson of this chapter, however, is that any theory that individuates functional types by appeal to capacities that led to some sort of historical success cannot account for the occurrence of malfunctions. The theory of selected functions fails to explicate in naturalistic terms the alleged normativity of functions, in which case the motivation for Preston's pluralism is not forthcoming. The instantiation view proposed by Griffiths (1993) is likewise deficient. Griffiths proposes that selected functions are a kind of systemic function. On his view, systemic functions generally are not normative, except those that arise in the course of being selected for; the process of selection for somehow transforms nonnormative systemic functions into normative ones. But the lesson of this chapter is that being selected for, insofar as it is a type of success, cannot confer a normative property upon systemic functions of any type. This is true, moreover, no matter when the relevant period of selection occurs. If it occurs in the past, present tokens lacking the property selected for have no functional standing; if it is occurring in the present, present tokens devoid of the efficacious property likewise have no functional standing. As a result, the "relational" view of functions—a variation on selected functions that allows for functions as a consequence of current selection (Walsh 1996)—fares no better than other versions of the theory in accounting for malfunctions. So, none of the specific versions of the combination approach—unification, pluralism, instantiation—can account for malfunctions. Insofar as advocates of the combination approach seek to account for malfunctions in terms of some kind of historical success, we must reject their theories.

VI Systemic Malfunctions

The argument of this chapter generalizes further still. The theory of systemic functions, on my view, cannot account for the occurrence of malfunctions. Of course, systemic functions are not defined in terms of historical success, but rather in terms of systemic capacities instantiated in systemic mechanisms. Moreover, the attribution of systemic functions is warranted so long as the capacity in fact exists in the system, so long as the underlying mechanisms that instantiate the capacity are intact. It is not necessary that the capacity is ever exercised or results in actual success; all that is required is the existence of the capacity. Nevertheless, when disease or damage results in the absence of the relevant systemic capacity, the corresponding systemic function likewise is absent. When one or more of the underlying mechanisms fail to develop, the associated systemic function fails to develop, and when one or more mechanisms is incapacitated, the systemic function is lost. Thus, although systemic functions are not defined by reference to historical success, they are defined in terms of capacities to contribute to the exercise of some higher-level capacity. They are defined in terms of the capacity for systemic success. Thus, when the capacity to contribute is absent, so too is the systemic function.

This, in fact, is the consensus: Systemic malfunctions are not possible. At least one theorist has claimed otherwise on the grounds that an incapacitated token nevertheless belongs to the systemic functional type. Thus, "[i]f a token of a component is not able to do whatever it is that other tokens do, . . . [if it is not able to perform those tasks that play] a distinguished role in the explanation of the capacities of the broader system, then that token component is malfunctional" (Godfrey-Smith 1993, 200). This is a reasonable suggestion only if there are compelling grounds for the claim that incapacitated tokens retain their membership in the functional type and not just the generic type. But there are no such grounds. To the contrary, systemic functional types are defined by reference to systemic capacities that contribute to some higher-level systemic capacity. Insofar as token traits lose or fail to acquire the defining capacity, they thereby lose or fail to acquire a place in the functional type. On what grounds other than possession of the defining capacity might they retain their membership in the functional type? Of course, we may wish to place the incapacitated tokens into a given functional type, but that indicates something about us, not about the nature of functions.

As suggested in the previous chapter, our inclination to place incapacitated tokens into functional types is plausibly an expression of our expectations regarding systemic functional types. We are inclined to see such tokens as members of a functional type because our experience with past instances or with an analogous type causes us to expect current tokens to exercise the relevant capacity. That we have such expectations is thus explicable. And that such expectations often facilitate inquiry is also easy to see. The world contains a wealth of natural traits distinguished by their systemic capacities; the world thus contains a wealth of systemic functional kinds. Some of these kinds exist and operate within systems that tend to persist over time. Organisms are exemplary. The greater the persistence and regularity of the kind, the stronger our generalizations concerning the kind and hence the stronger our expectations that future tokens will be behave in similar fashion. And, as suggested in chapters 4 and 6, conceptualizing natural traits in terms of systemic functions adds tractability to the process of inquiry. It enables us to anticipate and predict the behavior of complex systems in terms of systemic roles abstracted from the details of lower-level mechanisms. So, the regularity of hierarchical systems causes us to hold certain expectations concerning systemic functional types and the tractability that this adds to our inquiries only reinforces those expectations. When, therefore, an incapacitated token fails to satisfy our expectations, many of us feel as though it has somehow failed to satisfy a norm of evaluation that properly applies to such tokens. But this feeling is hardly decisive. We need only remind ourselves that the token possesses no such norm of evaluation and that we possess all manner of expectations informed by our knowledge of such systems. The temptation to attribute malfunctions to such tokens should, with time, fade away.

VII Conclusion

I conclude, then, that the historical approach and the combination approach to functions ought to be rejected. Both approaches fail to account for the possibility of malfunctions and hence both fail to account for what they insist is a central phenomenon that any such theory must explain. By contrast, I recommend the version of the theory of systemic functions developed and defended in these pages. This theory attempts to make a virtue out of the fact that it too cannot account for malfunctions. The basic thought is that functions are capacities that enable systems to accomplish the work that they do; and such functions exist within a system only if mechanisms within the system instantiate them. If the mechanisms are lost or somehow incapacitated, the systemic functions are lost as well. Our inclination to see the functions of natural traits as something grander than mere systemic capacities—as "proper" or "purposive" norms of nature—is not a datum that a theory of functions should explain, but rather a feature of our evolving conceptual scheme that should be explained away.

I believe we must accept that natural nonengineered traits do not possess the norms of performance that most theorists of functions wish to attribute to them. We must accept this or else admit that we are not naturalists after all. A virtue of the theory defended in this book is its acceptance of this fact and its attempts to explain away in psychological terms the sorts of inclinations that drive us toward the attribution of norms of performance. You may not be persuaded by the Humean speculations offered—they are, admittedly, the thinnest of speculations—but, as I have taken the care to emphasize, it is the general strategy that deserves consideration. We are indeed tempted to see the biological and psychological realms in terms of norms of performance; recall (from chapter 5) Darwin's inability to shake off the sense that the living realm is designed. But we are also committed to a general orientation toward inquiry that eschews the attribution of such norms to natural, nonengineered traits. In this context, the effort to explain away in Humean fashion our intuitions concerning the normative status of hearts, eyes, and thumbs is compelling and commendable. It is plausible that our temptation to attribute norms to token traits is an expression of our sentiments, including our more visceral and aesthetic responses to the natural realm. This general thought is plausible even if the details of my speculations are not.

We thus are driven toward a revisionist theory of functions. We should relinquish the claim that natural functions are normative in some respect or other. Our function attributions, insofar as they reflect our informed expectations, may depend upon our employment of various predictive or explanatory norms. But the functional traits themselves—the tokens possessed of the systemic capacities—are not the bearers of norms of any sort. They certainly are not the bearers of norms that persist when the requisite physical capacities do not exist. To think otherwise is, I suspect, to try to preserve within our naturalistic worldview a bit of old-fashioned metaphysical comfort disguised as science. But as Bishop Butler taught, everything is what it is. The attribution of norms to natural traits is a habit of mind—the habit of conceptualizing the living world in terms of design—we ought to break.

Appreciation and Acknowledgments

Several people helped me in the writing of this book, some directly by discussing or commenting on the ideas and arguments, others indirectly with encouragement or support of various sorts. I am deeply grateful to them all. My greatest debt is to Ann Cyptar, for love and beauty, and for unbounded support. Thanks, too, to Bill Lycan, for humor and learning, for abiding encouragement, and for being a generous long-distance colleague; to Bob Richardson, for persistent skepticism of philosophical theorizing carried on in the absence of scientific knowledge, for insightful feedback on several parts of the manuscript, and for warm hospitality during my tenure as a Taft postdoc at the University of Cincinnati; to George Harris, for inquiring after the general theoretical framework within which my views on functions acquire their larger significance, and for being a real colleague; to Robert Cummins, for unsurpassed work on functions and constructive criticisms on the entire manuscript; to Tom Adajian, for extended conversations on the very early versions of chapter 7, as well as countless other philosophical topics; to Jim Hopkins, Bill Lycan, Graham MacDonald, Karen Neander, Bob Richardson, Wayne Riggs, and Jay Rosenberg, for excellent criticisms of earlier versions of chapter 7; to David Buller, Robert Cummins, Bill Lycan, Richard Manning, Beth Preston, Bob Richardson, and Mark Risjord for helpful comments on an earlier version of chapter 3; and to George Harris and Jim Harris for helpful comments on parts of the penultimate draft.

I am also happily indebted to the National Endowment for the Humanities for a 1995 summer research stipend; to the Charles P. Taft Memorial Foundation for a 1996–97 research fellowship at the University of

Cincinnati, during which much of the groundwork for this book was done; to the philosophy department at UC for hospitality and intellectual stimulation; to the administration here at the College of William and Mary for generous financial support, including several summer research grants; and to the philosophy department at the College, for providing a one-semester leave from teaching duties during which much of this book was written.

Thanks to Kluwer Academic Publishers for kind permission to use my article "Malfunctions," published in *Biology and Philosophy* (2000), volume 15, pp. 19–38, and to Blackwell Publishers for kind permission to incorporate most of my article "The Nature of Natural Norms: Why Selected Functions Are Systemic Capacity Functions," published in *Noûs* (2000), volume 34, pp. 85–107.

References

Allen, C. & Bekoff, M. (1995) Biological Function, Adaptation, and Natural Design. *Philosophy of Science, 62,* 609–22.

Amundson, R. & Lauder, G. (1994). Function Without Purpose: The Uses of Causal Role Function in Evolutionary Biology. *Biology and Philosophy, 9,* 443–69.

Aristotle. (1984). Parts of Animals, *The Complete Works of Aristotle,* ed. Jonathan Barnes, trans. W. Ogle, pp. 994–1086. Princeton: Princeton University Press.

Ayala, F. (1970). Teleological Explanations in Evolutionary Biology. *Philosophy of Science, 37,* 1–15.

Barlow, N. (1958). *The Autobiography of Charles Darwin.* St. James's Place, London: Collins.

Bechtel, W. & Richardson, R. C. (1993). *Discovering Complexity: Decomposition and Localization as Strategies in Scientific Research.* Princeton: Princeton University Press.

Bedau, M. (1992). Where's the Good in Teleology? *Philosophy and Phenomenological Research, 52,* 781–805.

Bigelow, J. & Pargetter, R. (1987). Functions. *The Journal of Philosophy, 84,* 181–196.

Blackburn, S. (1998). *Ruling Passions: A Theory of Practical Reasoning.* Oxford: Oxford University Press.

Bock, W. J. & von Wahlert, G. (1965). Adaptation and the Form-Function Complex. *Evolution, 19,* 269–99.

Boorse, C. (1976). Wright on Functions. *Philosophical Review, 85,* 70–86.

Brandon, R. N. (1981). Biological Teleology: Questions and Explanations. *Studies in the History and Philosophy of Science, 12,* 91–105.

Brandon, R. N. (1982). The Levels of Selection. In *PSA 1980* (Vol. 2), ed. P. Asquith & R. Giere, pp. 427–39. East Lansing, MI: Philosophy of Science Association.

Brandon, R. N. (1990). *Adaptation and Environment.* Princeton: Princeton University Press.

Brandon, R. N. (1996). *Concepts and Methods in Evolutionary Biology.* New York: Cambridge University Press.

Brandon, R. N. (forthcoming). Teleology in Self-Organizing Systems.

Browne, J. (1995) *Charles Darwin Voyaging: Volume I of a Biography.* New York: Alfred A. Knopf.

Buller, D. (1998). Etiological Theories of Function: A Geographical Survey. *Biology and Philosophy, 13,* 505–27.

Buss, L. W. (1987). *The Evolution of Individuality.* Princeton: Princeton University Press.

Chase, W. & Simon, H. (1973). Perception in Chess. *Cognitive Psychology, 4,* 55–81.

Churchland, P. S. and Sejnowski, T. (1992). *The Computational Brain.* Cambridge: MIT Press.

Cosmides, L. & Tooby, J. (1987). From Evolution to Behavior: Evolutionary Psychology as the Missing Link. In *The Latest on the Best: Essays on Evolution and Optimality,* ed. J. Dupré, 277–306. Cambridge: MIT Press.

Cosmides, L. & Tooby, J. (1992). Cognitive Adaptations for Social Exchange. In *The Adapted Mind: Evolutionary Psychology and the Generation of Culture,* ed. J. H. Barkow, L. Cosmides & J. Tooby, 163–228. New York: Oxford University Press.

Cummins, R. (1975). Functional Analysis. *The Journal of Philosophy, 72,* 741–60.

Cummins, R. (1983). *The Nature of Psychological Explanation.* Cambridge: MIT Press.

Darwin, C. (1859). *On the Origin of Species (A Facsimile of the First Edition).* Cambridge: Harvard University Press.

Darwin, C. (1871). *The Descent of Man and Selection in Relation to Sex.* London: Murray.

Davies, P. S. (1994). Troubles for Direct Proper Functions. *Noûs, 28,* 363–81.

Davies, P. S. (1996). Discovering the Functional Mesh: On the Methods of Evolutionary Psychology. *Minds and Machines, 6,* 559–85.

Davies, P. S. (1999). The Conflict of Evolutionary Psychology. In *Where Biology Meets Psychology,* ed. V. G. Hardcastle. Cambridge: MIT Press.

Davies, P. S. (2000). Malfunctions. *Biology and Philosophy, 15,* 19–38.

Davies, P. S. (2000). The Nature of Natural Norms: Why Selected Functions Are Systemic Capacity Functions. *Noûs, 34,* 85–107.

Dawkins, R. (1976). *The Selfish Gene.* Oxford: Oxford University Press.

Dawkins, R. (1982). Replicators and Vehicles. In *Current Problems in Sociobiology*, ed. King's College Sociobiology Group, 45–64. Cambridge: Cambridge University Press.

Dennett, D. (1974). Why the Law of Effect Will Not Go Away. *Journal of the Theory of Social Behaviour, 5*, 169–87.

Dretske, F. (1995). *Naturalizing the Mind.* Cambridge: MIT Press.

Enç, B. (1979). Function Attributions and Functional Explanation. *Philosophy of Science, 46*, 343–65.

Enç, B. & Adams, F. (1992). Functions and Goal-Directedness. *Philosophy of Science, 59*, 635–54.

Field, H. (1980). *Science Without Numbers.* Oxford: Basil Blackwell.

Ghiselin, M. T. (1997). *Metaphysics and the Origin of Species.* Albany: State University of New York Press.

Gibbard, A. (1990). *Apt Feelings, Wise Choices: A Theory of Normative Judgment.* Cambridge: Harvard University Press.

Gillespie, N. C. (1979). *Charles Darwin and the Problem of Creation.* Chicago: University of Chicago Press.

Godfrey-Smith, P. (1993). Functions: Consensus Without Unity. *Pacific Philosophical Quarterly, 74*, 196–208.

Godfrey-Smith, P. (1994). A Modern History Theory of Functions. *Noûs, 28*, 344–62.

Gould, S. J. (1980). *The Panda's Thumb.* New York: Norton.

Gould, S. J. & Lewontin, R. C. (1979). The Spandrels of San Marco and the Panglossian Paradigm: A Critique of the Adaptationist Programme. *Proceedings of the Royal Society of London, B205*, 581–98.

Griffiths, P. E. (1992). Adaptive Explanations and the Concept of a Vestige. In *Trees of Life: Essays in Philosophy of Biology*, ed. P. E. Griffiths, 111–31. Dordrecht, Holland: Kluwer Academic Publishers.

Griffiths, P. E. (1993). Functional Analysis and Proper Functions. *British Journal for the Philosophy of Science, 44*, 409–22.

Griffiths, P. E. (1996). The Historical Turn in the Study of Adaptation. *British Journal for the Philosophy of Science, 47*, 511–532.

Griffiths, P. E. & Gray, R. D. (1994). Developmental Systems and Evolutionary Explanation. *The Journal of Philosophy, 91*, 277–304.

Hacking, I. (1983). *Representing and Intervening: Introductory Topics in the Philosophy of Science.* New York: Cambridge University Press.

Hagen, J. (1999). Retelling Experiments: H. B. D. Kettlewell's Studies of Industrial Melanism in Peppered Moths. *Biology and Philosophy, 14*, 39–54.

Harris, G. (1999). *Agent-Centered Morality: An Aristotelian Alternative to Kantian Internalism.* Berkeley and Los Angeles: University of California Press.

Haugeland, J. (1978). The Nature and Plausibility of Cognitivism. *Behavioral and Brain Sciences, 2,* 215–60.

Herschel, J. (1831). *Preliminary Discourse in the Study of Natural Philosophy.* London: Longman, Rees, Orme, Brown, and Green.

Hinton, G. (1986). Learning Distributed Representations of Concepts. In *Proceedings of the Eighth Annual Conference of the Cognitive Science Society.* Hillsdale, NJ: Lawrence Erlbaum.

Hull, D. (1973). *Darwin and His Critics: The Reception of Darwin's Theory of Evolution by the Scientific Community.* Cambridge: Harvard University Press.

Hull, D. (1981). Units of Evolution: A Metaphysical Essay. In *The Philosophy of Evolution,* ed. U. Jensen & R. Harre, 23–44. Brighton, UK: Harvester Press.

Hume, D. (1779). *Dialogues Concerning Natural Religion.* Indianapolis: Bobs-Merrill.

Kane, T. C. & Richardson, R. (1990). The Phenotype as the Level of Selection: Cave Organisms as Model Systems. *PSA 1990, 1,* 151–64.

Kettlewell, H. B. D. (1973). *The Evolution of Melanism: The Study of a Recurring Necessity.* Oxford: Oxford University Press.

Kingsolver, J. & Koehl, M. (1985). Aerodynamics, Thermoregulation, and the Evolution of Insect Wings: Differential Scaling and Evolutionary Change. *Evolution, 39,* 488–504.

Kitcher, P. (1981). Explanatory Unification. *Philosophy of Science, 48,* 507–31.

Kitcher, P. (1993). Function and Design. *Midwest Studies in Philosophy, XVIII,* 379–97.

Litchfield, H. (1915). *Emma Darwin: A Century of Family Letters* (2 Volumes). New York: D. Appleton.

Lloyd, E. A. (1988). *The Structure and Confirmation of Evolutionary Theory.* Princeton: Princeton University Press.

Lloyd, E. A. (1992). Unit of Selection. In *Keywords in Evolutionary Biology,* ed. E. F. Keller and E. A. Lloyd, 334–340. Cambridge: Harvard University Press.

Lycan, W. (1987). *Consciousness.* Cambridge: MIT Press.

Lycan, W. (1988). *Judgement and Justification.* New York: Cambridge University Press.

Lycan, W. (1990). (Ed.). *Mind and Cognition.* Cambridge: Basil Blackwell.

Matthen, M. (1988). Biological Functions and Perceptual Content. *The Journal of Philosophy, 85,* 5–27.

Mayr, E. (1961). Cause and Effect in Biology. *Science, 134,* 1501–6.

Mayr, E. (1963). *Animal Species and Evolution.* Cambridge: Harvard University Press.

McGinn, C. (1989). *Mental Content.* Oxford: Basil Blackwell.

McKitrick, M. (1993). Phylogenetic Constraint in Evolutionary Theory: Has it Any Explanatory Power? *Annual Review of Ecology and Systematics, 24,* 307–30.

Millikan, R. G. (1984). *Language, Thought, and Other Biological Categories: New Foundations for Realism.* Cambridge: MIT Press.

Millikan, R. G. (1989). In Defense of Proper Functions. *Philosophy of Science, 56,* 288–302.

Millikan, R. G. (1993). *White Queen Psychology and Other Essays for Alice.* Cambridge: MIT Press.

Millikan, R. G. (1999). Wings, Spoons, Pills, and Quills: A Pluralist Theory of Function. *The Journal of Philosophy, XCVI,* 191–206.

Mitchell, S. D. (1995). Function, Fitness, and Disposition. *Biology and Philosophy, 10,* 39–54.

Nagel, E. (1977). Teleology Revisited. *The Journal of Philosophy, 74,* 261–301.

Neander, K. (1991). Functions as Selected Effects: The Conceptual Analyst's Defense. *Philosophy of Science, 58,* 168–84.

Neander, K. (1995). Misrepresenting and Malfunctioning. *Philosophical Studies, 79,* 109–41.

Oyama, S. (1985). *The Ontology of Information.* Cambridge: Cambridge University Press.

Paley, W. (1802). *Natural Theology: Or Evidences of the Existence and Attributes of the Deity Collected from the Appearances of Naure.* London: reprinted Farnborough, Gregg, 1970.

Papineau, D. (1984). Representation and Explanation. *Philosophy of Science, 51,* 550–72.

Papineau, D. (1987). *Reality and Representation.* Oxford: Basil Blackwell.

Papineau, D. (1993). *Philosophical Naturalism.* Oxford: Blackwell Publishers.

Pinker, S. & Bloom, P. (1992). Natural Language and Natural Selection. In *The Adapted Mind: Evolutionary Psychology and the Generation of Culture,* ed. J. H. Barkow, L. Cosmides, & J. Tooby, 451–93. New York: Oxford University Press.

Plantinga, A. (1993). *Warrant and Proper Function.* New York: Oxford University Press.

Post, J. (1991). *Metaphysics: A Contemporary Introduction.* New York: Paragon House.

Preston, B. (1998). Why Is a Wing Like a Spoon? A Pluralist Theory of Functions. *The Journal of Philosophy, XCV,* 215–54.

Price, C. (1995). Functional Explanations and Natural Norms. *Ratio (New Series), 7,* 143–60.

Rowland, M. (1997). Teleological Semantics. *Mind, 106,* 279–303.

Rudge, D. W. (1999). Taking the Peppered Moth with a Grain of Salt. *Biology and Philosophy, 14,* 9–37.

Ruse, M. (1979). *The Darwinian Revolution: Science Red in Tooth and Claw.* Chicago: University of Chicago Press.

Salmon, W. (1971). *Statistical Explanation and Statistical Relevance.* Pittsburgh: University of Pittsburgh.

Schlosser, G. (1998). Self-Re-Production and Functionality: A Systems-Theoretical Approach to Teleological Explanation. *Synthese, 116,* 303–354.

Selverston, A. (1988). A Consideration of Invertebrate Central Pattern Generators as Computational Data Bases. *Neural Networks, 1,* 109–117.

Shapiro, L. A. (1998). Do's and Don't's for Darwinizing Psychology. In *The Evolution of Mind,* ed. D. D. Cummins & C. Allen, 243–59. New York: Oxford University Press.

Simon, H. (1969). *The Sciences of the Artificial.* Cambridge: MIT Press.

Simon, H. (1973). The Organization of Complex Systems. In *Hierarchy Theory: The Challenge of Complex Systems,* ed. H. Pattee, 1–27. New York: Braziller.

Sober, E. (1984). *The Nature of Selection: Evolutionary Theory in Philosophical Focus.* Cambridge: MIT Press.

Sober, E. (1990). The Poverty of Pluralism: A Reply to Sterelny and Kitcher. *The Journal of Philosophy, 87,* 151–58.

Sober, E. (1994). *Conceptual Issues in Evolutionary Biology.* Cambridge: MIT Press.

Sober, E. & Wilson, D. S. (1994). A Critical Review of Philosophical Work on the Units of Selection Problem. *Philosophy of Science, 61,* 534–55.

Sober, E. & Wilson, D. S. (1998). *Unto Others: The Evolution and Psychology of Unselfish Behavior.* Cambridge: Harvard University Press.

Sterelny, K. & Kitcher, P. (1988). The Return of the Gene. *The Journal of Philosophy, 85,* 339–61.

Suppe, F. (1979). Theory Structure. In *Current Research in Philosophy of Science,* 317–38. East Lansing, MI: Philosophy of Science Association.

Suppes, P. (1967). What Is a Scientific Theory? In *Philosophy of Science Today,* ed. S. Morgenbesser, 55–67. New York: Meridian.

Thompson, P. (1988). *The Structure of Biological Theories.* Albany: State University of New York Press.

Tinbergen, N. (1963). On the Aims and Methods of Ethology. *Zeitschrift Für Tierpsychologie, 20,* 410–29.

van Fraassen, B. (1972). A Formal Approach to the Philosophy of Science. In *Paradigms and Paradoxes,* ed. R. Colodny, 303–66. Pittsburgh: University of Pittsburgh Press.

van Fraassen, B. (1980). *The Scientific Image.* Oxford: Oxford University Press.

Walsh, D. M. (1996). Fitness and Function. *British Journal for the Philosophy of Science, 47,* 553–74.

Walsh, D. M. & Ariew, A. (1996). A Taxonomy of Functions. *Canadian Journal of Philosophy, 26,* 493–514.

Watson, J. D. (1970). *Molecular Biology of the Gene.* New York: W. A. Benjamin.

Wimsatt, W. (1972). Teleology and the Logical Structure of Function Statements. *Studies in the History and Philosophy of Science, 3,* 1–80.

Wimsatt, W. (1986). Forms of Aggregativity. In *Human Nature and Natural Knowledge,* ed. A. Donagan, A. Perovich, & M. Wedin, 259–91. Dordrecht, Holland: D. Reidel.

Wimsatt, W. (1997). Functional Organization, Functional Analogy, and Functional Inference. *Evolution and Cognition,* 3, 102–32.

Woodfield, A. (1976). *Teleology.* Cambridge: Cambridge University Press.

Wright, L. (1973). Functions. *Philosophical Review, 82,* 139–68.

Wright, L. (1976). *Teleological Explanations: An Etiological Analysis of Goals and Functions.* Berkeley and Los Angeles: University of California Press.

Index